Stochastic Water Demand Modelling

KWR Watercycle Research Institute Series

Stochastic Water Demand Modelling
Hydraulics in Water Distribution Networks

Mirjam Blokker

IWA Publishing
London • New York

Published by **IWA Publishing**
Alliance House
12 Caxton Street
London SW1H 0QS, UK
Telephone: +44 (0)20 7654 5500
Fax: +44 (0)20 7654 5555
Email: publications@iwap.co.uk
Web: www.iwapublishing.com

First published 2011

British Library Cataloguing in Publication Data
A CIP catalogue record for this book is available from the British Library

Library of Congress Cataloging-in-Publication Data
A catalog record for this book is available from the Library of Congress

ISBN13: 9781780400280
ISBN: 1780400284

Contents

Chapter 1

Introduction

1.1 WATER DEMAND MODELLING

Drinking water companies are interested in their customers' water demands. One reason is that it is good practice to charge customers for the amount of drinking water they consume. Another reason is that drinking water companies want to accurately predict and secure future water supply. Understanding water demand can either be used to predict future water use or to influence future water use; usually by encouraging people to save water. Future water supply can be the demand within the next hour, or tomorrow, or as long term as the water supply 20 years from now. To date, knowledge of water demand has been largely based on water flow measurements from residential water meters and flow meters at Treatment works and pumping stations. As a result, demands are studied, and even modelled, on a relatively large temporal scale (usually one hour to one day), and large spatial scale (usually the level of a district metered area or supply area).

This book is concerned with a more detailed type of demand modelling. To this end, the water demand model SIMDEUM (**SIM**ulation of water **D**emand; an **E**nd-**U**se **M**odel) was developed. With a one second time step and at the fixture level in households, this model is more detailed in both temporal and spatial scales. The application of this detailed model is not specifically related to billing, nor to securing the quantity of water supply. In fact, it is related to securing the quality of water supply at the customers' tap.

Drinking water companies strive to produce high quality drinking water. However, this quality can deteriorate in the drinking water distribution systems (DWDS). Water quality deterioration in the DWDS occurs through a variety of processes, such as contaminant propagation, residual disinfectant decay, disinfection by-product formation, biofilm formation, taste and odour development, corrosion, and particle accumulation and mobilisation. These processes are influenced by a number of factors such as contact time with the pipe wall or biofilm, temperature, shear stresses at the pipe wall and interaction with particles and biofilm on the wall. Furthermore, all of these water quality processes are influenced by the hydraulic conditions in the network. Contact time with the pipe wall or biofilm is influenced by the residence time; the shear stress is influenced by the velocity; the temperature in the DWDS is influenced by the soil temperature, residence time and flow velocity. The propagation of dissolved substances is not only influenced by advection, and therefore average velocities, but also by dispersion, and therefore stagnant and laminar flows affect the propagation. The water quality deterioration in the DWDS is therefore closely related to hydraulic processes, even down to the level of instantaneous or very localised hydraulic changes. The water demand model SIMDEUM was developed to gain further insight into the hydraulic processes experienced within water distribution networks, and more specifically, to investigate stagnant, laminar and turbulent flows, flow direction reversals, residence times and instantaneous flow velocities. With a better understanding of the hydraulics, further insight can be gained into water quality processes in the water distribution networks.

1.2 RESEARCH OBJECTIVES

The following research objectives were defined:

The first objective was to develop the water demand model SIMDEUM and to validate its ability to predict (maximum) water demand for several types of residential and non-residential users.

The second objective was to apply SIMDEUM in case studies related to the modelling of water quality in the DWDS. The focus was on hydraulic conditions which affect water quality processes in the DWDS, such as residence time and maximum flow velocity.

The last research objective was to evaluate the added value of SIMDEUM in hydraulic and water quality network modelling.

1.3 OUTLINE

Chapter 2 investigates how a network model would benefit from a different approach of demand modelling. A literature review and examination of the characteristics of actual demands show that modelling the water quality in the DWDS requires a more detailed demand model. It demonstrates the potential added value of a detailed demand model such as SIMDEUM in water quality modelling. It concludes that field validation of this type of demand modelling is necessary. Chapters 6 and 7 respond to that requirement.

Chapter 3 describes how SIMDEUM works and it verifies how the model predicts (maximum) water demand for Dutch residences. Chapter 4 shows that the model can be extended beyond residential water demands; it verifies how the model predicts water demands in offices, hotels and nursing homes. In Chapter 5, it was also confirmed that SIMDEUM can be applied to simulate the water demand of US households. This chapter also compares the approach of SIMDEUM end-use modelling to the approach of an existing detailed demand model (the PRP model).

Next, the demand patterns from SIMDEUM were applied in network models in three case studies. Chapter 6 describes the application of SIMDEUM demand patterns in a small network model (140 residences) and an intermediate-scaled network model (1000 residences, 3 hotels and some beach clubs). In these networks, the flow into the network and the residence times were measured by means of a tracer study. Residence times from a network model with a new approach of demand allocation (bottom-up) were compared to the results of the network model using a conventional way of demand allocation (top-down) and to the measured residence times. The third case study (Chapter 7) investigated the relationship between maximum flow velocities and discolouration risk. In a real DWDS, the maximum flow velocities were established with a network model plus SIMDEUM demand patterns. The discolouration risk was determined by measuring turbidity generated by mobilised particles during flushing.

Chapter 8 provides a general discussion. It further evaluates the added value of SIMDEUM in distribution network models and discusses suggested future research and applications.

Appendix A elaborates on the statistics that were used in the book, including the parameters that were used to compare measurements and models and to determine the goodness-of-fit. It explains auto- and cross-correlation. It also explains the probability distribution functions that were used.

Chapter 2

Importance of demand modelling in network water quality models: a review*

ABSTRACT

Today, there is a growing interest in network water quality modelling. The water quality issues of interest relate to both dissolved and particulate substances. For dissolved substances, the main interest is in residual chlorine and (microbiological) contaminant propagation; for particulate substances it is in particles leading to discolouration. There is a strong influence of flows and velocities on transport, mixing, production and decay of these substances in the network. Thus, the hydraulics should be modelled on a relevant temporal and spatial scale. This imposes a different approach to demand modelling which is reviewed in this article.

For the large diameter pipes that comprise the transport portion of a typical municipal pipe system, a skeletonised network model with a top-down approach of demand pattern allocation, a hydraulic time step of one hour, and a pure advection-reaction water quality model will usually suffice. For the smaller diameter pipes that comprise the distribution portion of a municipal pipe system, an all-pipes network model with a bottom-up approach of demand pattern allocation, a hydraulic time step of one minute or less, and a water quality model that considers dispersion and transients may be needed.

Demand models that provide stochastic residential demands per individual home and on a one-second time scale are available. A stochastic demand-based network water quality model needs to be developed and validated with field measurements. Such a model will be probabilistic in nature and will offer a new perspective for assessing water quality in the drinking water distribution system.

2.1 INTRODUCTION

The goal of drinking water companies is to supply their customers with good quality drinking water 24 hours per day. With respect to water quality, the focus has for many years been on the drinking water treatment. Recently, interest in water quality in the drinking water distribution system (DWDS) has been growing. On the one hand, this is driven by customers who expect the water company to ensure the best water quality by preventing such obvious deficiencies in water quality as discolouration and (in many countries) by assuring a sufficient level of chlorine residual. On the other hand, since '9/11' there is a growing concern about

*Reprinted with adaptations from Blokker, E. J. M., Vreeburg, J. H. G., Buchberger, S. G., and van Dijk, J. C. (2008). "Importance of demand modelling in network water quality models: a review." *Drink. Water Eng. Sci.*, 1(1), 27–38.

(deliberate) contaminations in the DWDS. Consequently, there is an interest in the behaviour of both particulate and dissolved substances throughout the DWDS (Powell et al. 2004).

In this chapter, transport mains are defined as pipes that typically do not supply customers directly; customer connections are attached only to distribution mains (Figure 2.1). Transport mains have relatively large diameters and supply distribution mains. As a result, transport mains have only a few demand nodes, with demands that show a high cross-correlation (i.e. all demand nodes show a similar demand profile over the day), the flows show a high auto-correlation at a time sacle of 5 to 15 minutes (i.e. the flows are relatively constant, do not vary rapidly) and are mainly turbulent with typical maximum velocities of 0.5–1.0 m/s (Vreeburg 2010). As a consequence of the high velocities and the fact that no customers are directly connected to the transport network, there is a low discolouration risk in transport mains. A transport network, therefore, requires only a relatively simple hydraulic model (e.g. EPANET) which can be constructed from basic pipe information (diameters, lengths and pipe material) and driven by strongly correlated demand profiles applied to nodes. The model is typically calibrated with pressure measurements (Kapelan 2002).

Figure 2.1 Part of a distribution network. The line colour and thickness represent the diameter of the pipes, the circles are demand nodes, open circles are nodes with zero demand. The thick dark lines are typically mains with a transport function (i.e. large diameters and very few demand nodes that are directly connected to it); the thin light coloured lines are mains with a distribution function (they supply to customers)

Distribution mains have many demand nodes, and instantaneous demands among individual homes show little auto- and cross-correlation (Filion et al. 2006). A distribution network is usually designed for fire flow demands that are typically much higher than domestic demands (Vreeburg 2010). Therefore, under normal operating conditions, the maximum velocities in distribution mains can be very low (smaller than 0.01 m/s) and change rapidly. Flow directions may reverse and travel times may be as long as 100 hours due to stagnation (Blokker et al. 2006; Buchberger et al. 2003). In the distribution portion of the network, particles do accumulate and are mobilised (Blokker et al. 2009b; Vreeburg 2010). This means a distribution mains model may need a rather complex structure for demand allocation.

In modelling water quality in the DWDS, the essential aspects are transport, mixing, production and decay. Sediment behaviour, and thus discolouration risk, in a DWDS is strongly related to hydraulics (Slaats et al. 2003; Vreeburg 2010). The spread of dissolved contaminants through the DWDS is strongly related to the flows through the network (Grayman et al. 2006). The current water quality models are only validated for the transport network. Since consumers are located in the distribution part of the network, a water quality model at that level is important. Because flows are more variable at the periphery of the DWDS, water quality models at this level may require a different approach than the models currently available.

The key element for a water quality model for a DWDS is an accurate hydraulic model and, therefore, detailed knowledge of water demands is essential. This chapter reviews the influence of the (stochastic) demands on water quality models and the consequential constraints on demand modelling. First is a review of water quality modelling of dissolved matter and its relation to demands. Next, water quality modelling of particulate matter and its relation to hydraulic conditions is described. Thirdly, the characteristics of demands in hydraulic network models and in network water quality models are discussed.

2.2 WATER QUALITY MODELLING – DISSOLVED MATTER

With increasing computational power, hydraulic network models are used more and more for water quality related subjects, such as determining residual chlorine (Bowden et al. 2006; Propato and Uber 2004) and disinfection by-products in the DWDS (under the US EPA Stage 2 Disinfection By-Products Rule (USEPA 2006)), optimum sensor placement for detection of biological and chemical contaminations (Berry et al. 2005; Nilsson et al. 2005) and source location inversion after a contaminant is detected (McKenna et al. 2005).

Water quality in a network model can be described with the Advection-Dispersion-Reaction (ADR) equation:

$$\frac{\partial C}{\partial t} + u\frac{\partial C}{\partial x} = E\frac{\partial^2 C}{\partial x^2} + f(C) \tag{2.1}$$

where C is the cross-sectional average concentration (the water quality parameter, usually in mg/L), t is the time (s), u is the mean flow velocity (m/s), x is the direction of the flow, E represents the mixing (axial dispersion) coefficient in one-dimensional flow (m^2/s) and $f(C)$ is a reaction function. Dispersion in this case is the Taylor dispersion, which is an effect in fluid mechanics in which the flow induced by a force gradient (caused by the difference in velocity at the center and at the wall of the pipe) can increase the effective diffusivity of a solute in water. Effectively, it causes mixing and longitudinal spreading of a solute. The left-hand term of this equation depicts the advection and mainly depends on bulk movement of the water. The first term on the right-hand side depicts the dispersion and the last term represents the reaction; both terms on the right-hand side of Eq. 2.1 depend on the type and nature of the considered substance. The reaction function can be very diverse for different substances. In most instances, however, a simple first-order reaction is assumed, e.g. for chlorine decay, $f(C) = -KC$ with K the reaction constant. The reaction function can include a production term.

The hydraulic network solver EPANET (Rossman 2000) comes with a water quality module, as do many commercially available network analysis programmes. The water quality module enables the user to calculate travel times and to model the migration of a tracer (both conservative and non-conservative) through a network. It models advection and reaction with the pipe wall and the bulk of the water, but it does not take dispersion into account (i.e. neglects the first term on the right-hand side of the ADR equation). While EPANET can handle many different time scales (i.e. time intervals over which demands are time averaged), a time scale of one hour is commonly used. The solver assumes that the network is well defined (known pipe diameter, pipe roughness and network layout), that demands are known, and that water quality reactions (under the influence of residence times and interaction with the pipe wall) are known. Furthermore, EPANET assumes perfect mixing at junctions and steady-state hydraulic conditions during every computational interval. Hence, EPANET is not suitable for simulating transient flow in pipe networks. The accuracy of the calculated results depends on the validity of these assumptions.

To further the water quality models, research is done on several of the assumptions in the models. In this review, the focus is on model deficiencies with respect to flows and velocities.

With respect to advection, Eq. 2.1 shows that time, and thus travel time, is an important factor, as is the velocity of the water. A proper assumption of demands is a key factor in solving Eq. 2.1. Several authors (Filion et al. 2005; McKenna et al. 2005; Pasha and Lansey 2010) have shown the importance of uncertainty of demands in water quality models. The needed detail in demand allocation is yet unknown.

Advection is also related to mixing. The conventional assumption of perfect mixing at junctions has been studied with measurements and Computational Fluid Dynamics modelling (Austin et al. 2008; Romero-Gomez et al. 2008a). The studies showed that at T-junctions, that are at least a few pipe diameters apart, perfect mixing can be assumed, while in cross junctions less than 10% mixing may occur. In fact, at cross junctions the rate of mixture in the two outgoing arms depends on the Reynolds numbers (and thus the flow rates) in the two incoming arms.

When looking at smaller time steps, a steady-state assumption may not be valid. Karney et al. (2006) investigated the modelling of unsteadiness in flow conditions with several mathematical models such as extended period approaches (like EPANET does), a rigid water column model that includes inertia effects, and a water hammer model that includes small compressibility effects. The time scale of boundary and flow adjustments relative to the water hammer time scale were found to be important for characterising the system response and judging the unsteadiness in a system. When for certain applications the required time step would be shorter than several minutes, the impact of taking inertia and compressibility into account should be studied further.

The dispersion term in Eq. 2.1 is small in the case of turbulent flow, but cannot be neglected in the case of laminar flow. Gill and Sankarasubramanian (1970) derived an exact but cumbersome expression showing that the *instantaneous* rate of dispersion in fully-developed steady laminar flow grows with time and asymptotically approaches the equilibrium dispersion rate E_T given by Taylor (1953),

$$E_T = \frac{d^2u^2}{192D} \tag{2.2}$$

where D is the molecular diffusivity of a solute (m^2/s) and d is the pipe diameter (m). Lee (2004) simplified the 1970 G&S result and provided a theoretical approximation for the *time-averaged* unsteady rate of dispersion, $\overline{E(t)}$, for a solute moving in steady laminar flow through a pipe,

$$\overline{E(t)} = E_T\left[1 - \frac{1 - \exp[-16T(t)]}{16T(t)}\right] \tag{2.3}$$

Here, $T(t) = 4Dt/d^2$ is dimensionless Taylor time and t represents the mean travel time through the pipe. When Taylor time is large, Eq. 2.3 reduces to Eq. 2.2. For nearly all network links, however, Taylor time is very small [e.g. $T(t) < 0.01$]. In this case, the expression in Eq. 2.3 can be further simplified,

$$\overline{E(t)} \approx \frac{u^2 t}{6} = \frac{ul}{6} \tag{2.4}$$

where l is the length of the pipe section (m). To illustrate, consider a solute with diffusivity $D = 10^{-9}$ m^2/s transported in steady fully developed laminar flow (say Re = 1000) at 20°C through a pipe with $d = 0.15$ m and $l = 100$ m. The corresponding average velocity is $u = 6.7 \times 10^{-3}$ m/s. Hence, the mean travel time through the pipe link is $t = l/u = 15{,}000$ s and the corresponding dimensionless Taylor time is $T(t = 15{,}000\ s) = 0.0027$. For this condition, Eqs. 2.3 and 2.4 give similar results, namely, $\overline{E(t)} = 0.1105$ m^2/s and 0.1117 m^2/s, respectively. These estimates of the dispersion rate are eight orders of magnitude greater than the rate of molecular diffusivity. However, they are only two percent of the equilibrium value given by Taylor's formula in Eq. 2.2, $E_T = 5.26$ m^2/s. Owing to small molecular diffusivity and relatively large pipe diameters, it is virtually impossible in real water distribution systems for the time-averaged rate of laminar dispersion to attain the equilibrium value given in Eq. 2.2.

Recent preliminary experimental evidence indicates that Eqs. 2.3 and 2.4 tend to slightly over-estimate the actual time-averaged rate of dispersion observed in controlled laboratory runs (Romero-Gomez et al. 2008b). The reasons for this discrepancy are not clear and this is the subject of ongoing research investigations.

The influence of dispersion in water quality modelling was tested with (two-dimensional) ADR models (Li 2006; Tzatchkov et al. 2002). Li (2006) showed that dispersion is important in laminar flows and thus especially in the parts of DWDS that have pipe diameters designed for fire flows but with small normal flows. Dispersion is not directly affected by flow pattern or time scale, although the tests of Romero-Gomez et al. (2008b) seem to suggest that the dispersion coefficient is related to the Reynolds number. Flow pattern and time scale do, however, affect the probability of stagnation, laminar and turbulent flows, and thus indirectly do have an effect on dispersion.

Powell et al. (2004) have established that there is a need to further investigate the reaction parameters for chlorine decay, disinfectant by-products and bacterial regrowth. Where the reaction constant K involves a reaction with the pipe wall, the stagnation time is of importance. The flow regime (laminar or turbulent flow) and thus flow velocities are important as they affect chlorine decay rates (Menaia et al. 2003).

2.3 WATER QUALITY MODELLING – PARTICULATE MATTER

Discolouration is believed to be caused by accumulated particles that are mobilised during moments of high flow (Vreeburg and Boxall 2007). Vreeburg (2010) developed the conceptual model that particles, originating from the treatment plant, are, under normal flow conditions, regularly deposited on the pipe wall and resuspended. Incidental high flows (e.g. when a fire hydrant is opened), can mobilise the particles on the pipe wall and can thus lead to discoloured water (Figure 2.2). Vreeburg (2010) has shown that the discolouration risk can be reduced with three types of measures: the first is to prevent material from entering the DWDS by optimising the water treatment; the second is to prevent particles from accumulating in the DWDS by designing self-cleaning networks (Vreeburg 2010); and, the third is to remove particles by cleaning (flushing) the DWDS in a timely manner. Although these three approaches have proven to reduce the discolouration risk, the exact relationship between the hydraulics and particles behaviour (under what conditions do they accumulate and mobilise?) is still unknown. More insight into the hydraulic conditions can further support these measures.

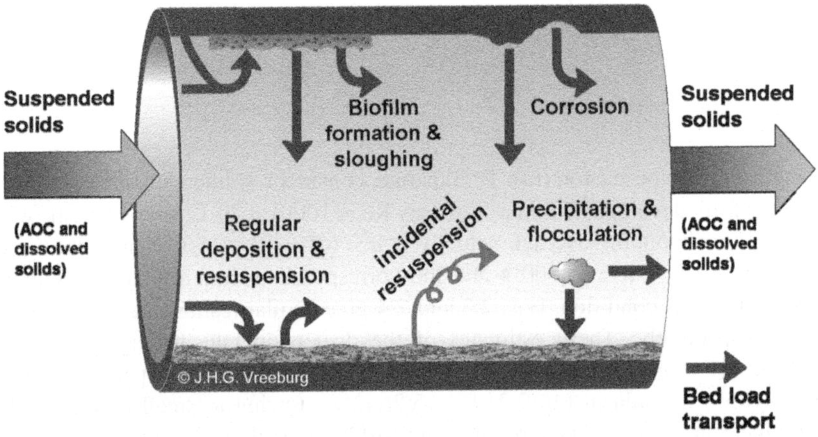

Figure 2.2 Processes related to particles in the distribution network (Vreeburg 2010)

Boxall et al. (2001) suggested that the size range of the particles related to discolouration was predominately less than 0.050 mm. They showed that it is unlikely that gravitational settling alone will be a sufficient force for accumulation of such small particles as turbulent forces generated by even the lowest flows within a distribution system are likely to be sufficient to overcome gravity settling forces. Vreeburg and Boxall (2007) suggest that turbophoresis could play a role. This means particles are transported from the bulk fluid to less turbid regions near the wall where they can be trapped in cohesive layers.

Husband et al. (2008) did discolouration tests in the laboratory with tap water that had a high natural iron concentration. They not only tested with constant flow velocities, but also with a dynamic flow pattern, of which the flow was changed every 15 minutes. The tests suggest that sediment settling and resuspension are affected differently by dynamic flows and constant flows. At constant flows, no matter how high they are, pipes may foul. A high velocity that occurs only seldom may be able to resuspend the sediment instantaneously and keep the pipe clean.

In a CRC Research report, Ryan et al. (2008) have described their Particle Sediment Model (PSM). Based on lab measurements, they have modelled particle accumulation through both gravitational settling and particle wall deposition. The physics behind the particle wall deposition model were not explained. Ryan et al. (2008) also modelled particle mobilisation through a relation with the velocity.

Vreeburg and Boxall (2007) concluded that the mechanisms leading to discolouration events are complex and poorly understood. Their basic concept of the cause of discolouration is that particles are attached by some means to the pipe wall. In normal flow the particles stay in their place and do not affect the aesthetic quality of the water. If flows are increased above normal, scouring forces and shear stress increase consequently and then the particles may be mobilised, sometimes leading to customer complaints.

Self-cleaning distribution networks (Vreeburg's approach 2) are supposed to be effective because a regularly occurring threshold velocity prevents particles from accumulating in the network. The threshold design velocity for self-cleaning DWDS is set to 0.4 m/s. The demand at which this threshold value occurs is determined with the so-called $q\sqrt{n}$ method (Vreeburg, 2007). Field measurements in the Netherlands have shown that the self-cleaning concept is feasible in real networks (Blokker et al. 2009b). Measurements with particle counters and flushing tests showed that branched systems with small diameters are cleaner (i.e. less sediment has accumulated) than conventional networks that were designed for fire flow demands. The field measurements also showed that the current method used to estimate the maximum flow (the $q\sqrt{n}$ method) overestimates the regularly occurring flow, meaning that the flow for which the DWDS was designed (almost) never takes place. Thus, the networks appear to be self-cleaning at velocities below 0.4 m/s. The actual velocity at which sediment resuspends is therefore smaller than the design velocity of 0.4 m/s. However, it is unclear what the actual velocity is at which sediment resuspends and how often this velocity has to occur in order for the DWDS to be self-cleaning. Also, there is a need for a more accurate method to estimate the daily demand.

Laboratory tests in the Netherlands (Slaats et al. 2003) have shown that sediment substitutes (sand, iron oxide and flour) of different sizes and densities are partly mobilised at velocities of 0.1 to 0.15 m/s and fully mobilised at velocities of 0.15 to 0.25 m/s. Ackers et al. (2001) found that (gravity settled) sediment with realistic diameters (45 μm) and a relatively high specific density (2600–3100 kg/m^3), started moving at flow velocities of 0.2 to 0.25 m/s. Ryan et al. (2008) have researched the velocites at which particles accumulated and were mobilised; they particles that were obtained from flushing actions throughout Australia. Mobilisation was found to occur at 0.2 to 0.3 m/s. Particles depositing to the wall was found at a range of (constant) velocities; at a velocity of 0.2 m/s it could take several hours to a few days before all sediment had attached to the pipe wall. It is unclear which flushing velocities were used to collect the particles and thus it is unclear if all particles were removed from the water mains or only the particles that could easily be mobilised. It is unclear from their study whether the matterial that would only be mobilised at a higher velocity was missed in their test.

Some attempts have been undertaken to model particle behaviour in water distribution networks. Boxall and Saul (2005) have developed a 'predictor of discolouration events in distribution systems' (PODDS). This model is based on the assumption that normal hydraulics forces (i.e. maximum daily shear stress) condition the material layer strength and hence control the discolouration potential (or discolouration risk).

Ryan et al. (2008) have implemented their PSM model onto EPANET. PSM predicts where in a network the most particles will accumulate and could thus indicate where flushing is required most. Their starting point is the ADR model; the reaction function for particles includes a velocity term. In the gravitational settling model a particle cloud is assumed, defined by a non-dimensional particle cloud height s (proportional to the pipe diameter). When all particles are settled, $s = 0$. When the flow velocity is larger than a certain threshold velocity (u_{rs}, the resuspension velocity), all particles are brought in suspension and $s = 1$. When the flow velocity is smaller than a certain threshold velocity (u_d, the deposition velocity), particles settle with a downward velocity (u_s, the settling velocity) and $0 < s < 1$:

$$s(t + \Delta t) = s(t) - \frac{u_s \Delta t}{d}, \quad \text{for } u < u_d \tag{2.5}$$

$$s(t) = 1, \quad \text{for } u \geq u_{rs} \tag{2.6}$$

In the particle wall deposition model (Ryan et al. 2008), the concentration of particles in suspension (C, in mg/L) is described as follows:

$$\frac{\partial C}{\partial t} = -\alpha(C - C_\infty) \tag{2.7}$$

Here, α is a decay coefficient (s^{-1}) and C_∞ is the final steady-state concentration of particles (mg/L). $C_\infty = \beta C_w$, with C_w the mass of particles on the wall per unit volume of water (mg/L), and β the dimensionless wall mass coefficient. The value of the parameter β depends on the flow velocity.

A test of PSM on a Dutch distribution network and a Dutch transport main showed that this type of model can work (Vogelaar and Blokker 2010). However, the wall deposition model does not seem to be suitable for variable flow velocities. The equilibrium of particles in suspension and attached to the wall depends on the flow velocity and there are more particles on the wall at higher velocities. The result of using only Eq. 2.7 for this mechanism is that at decreasing velocities particles return to the water phase without any hydraulic disturbance.

The self-cleaning design principles have mainly been applied to the peripheral zones of the distribution system which can be laid out as branched networks (sections of up to 250 residential connections). Even though the $q\sqrt{n}$ method overestimates the flows and the design velocity of 0.4 m/s might be a conservative value, the combination of these rules leads to self-cleaning networks (Blokker et al. 2009b). In order to scale-up the self-cleaning principle to the rest of the (looped) network, it is important to look into a better estimate of the regular occurring maximum flows because the $q\sqrt{n}$ method cannot easily be applied in looped networks. Buchberger et al. (2008) have used the PRP model (Buchberger et al. 2003) to derive that the maximum flow equals '$k_1 n + k_2\sqrt{n}$', with n the number of homes and the constants k_1 and k_2 are related to the PRP parameters (see Sect. 2.5). Since particle behaviour is related to instantaneous (peak) flows, modelling of particles in the network requires short time scales. More research must be done on the relationship between hydraulics and particles mobilisation (i.e. establish the actual self-cleaning velocity).

2.4 DEMANDS IN HYDRAULIC NETWORK MODELS

Demand modelling is done on different temporal and spatial aggregation levels, depending on the model's purpose. Three different levels of demand modelling and consequently network modelling can be distinguished. The highest level is for planning the operation of the treatment plant, for which it is important to model the demand per day for the total supply area of a pumping station. The second level is modelling the transport level or the level to which the assumption of cross-correlation is still sufficient, while for water quality modelling on a distribution level (the third level) a time scale on the order of minutes may be important (Blokker et al. 2006; Li and Buchberger 2004). Even at the transport level, skeletonising the network model and aggregation of demands can only be done to a limited extent without getting errors in the water quality model results, in this case the chlorine concentration at the nodes (Saldarriaga et al. 2009).

Temporal and spatial aggregation of demands is related to cross- and auto-correlation of flows. A high cross-correlation means that demand patterns at different nodes are similar (flows are proportional to each other). A high auto-correlation is found when flow patterns change gradually. Cross- and auto-correlation thus are related to maximum flow rates and the stagnation time. This does not only influence water quality; the amount of cross-correlation is important with respect to the reliability of a DWDS regarding nodal demands (Filion et al. 2005) and thus the cost (Babayan et al. 2005); auto-correlation is important with respect to the resilience of a DWDS, i.e. the time to restore service after a break (Filion et al. 2005). Several authors (Filion et al. 2006; Li and Buchberger 2007; Moughton et al. 2006) have looked at the effect of temporal and spatial aggregation of demands on cross- and auto-correlation. They have shown that the longer the time scale and the higher the aggregation level, the higher the (cross-) correlation. When looking at time scales of 1 hour and demand nodes that represent 10 or more connections, the assumption of cross-correlation is valid. This means that strongly correlated demand patterns can be applied in the hydraulic model.

Figure 2.3 and 2.4 show the mean and variance ($\mu \pm \sigma$, representing the 70% confidence interval and $\mu \pm 2\sigma$, the 95% confidence interval) of cross- and lag-1 auto-correlation coefficient for different time scales (1 to 60 min) and spatial scales (1, 10 and 20 homes per demand node) of 50 demand patterns as were measured in 1997 in 21 homes in Milford, Ohio (Buchberger et al. 2003). It shows that the cross-correlation for demand patterns of individual homes or at short time steps are low (the lower bound of the 95% confidence interval is not above 0) and that only for 20 homes and 15 min, the lower bound of the 95% confidence interval of the cross-correlation is above 20%. The lag-1 auto-correlation coefficient for short time steps can be high due to the high number of instances of zero flow. With an increasing time step, the lag-1 auto-correlation coefficient at first decreases with a decrease in zero flow instances and then increases with longer time steps, which is related to a more gradually changing pattern. For individual homes the lag-1 auto-correlation coefficient is low (the lower bound of the 95% confidence interval is not above 0) due to the stochastic nature of the water use. For 10 homes or more, the average lag-1 auto-correlation coefficient is stable at a time step of about 15 minutes or more, based on data from the Milford field study.

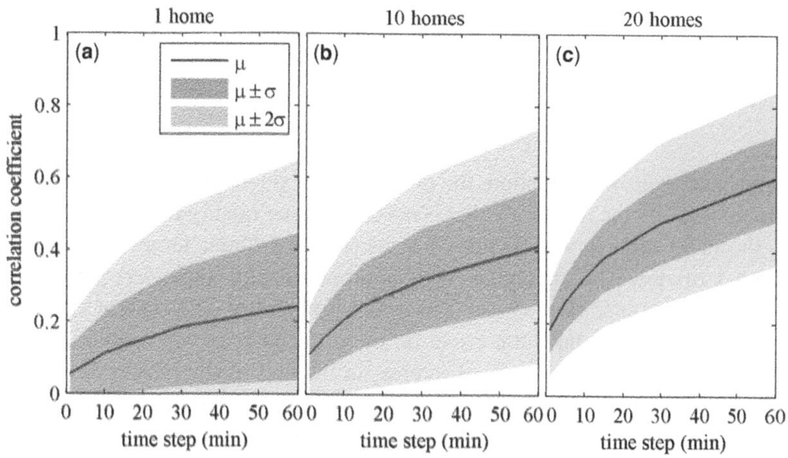

Figure 2.3 Mean (μ) and variation (σ) of cross-correlation of measured patterns (Milford, OH; Buchberger et al. 2003) on different temporal and spatial scales; a) 1, b) 10 and c) 20 homes

In a preliminary study Tzatchkov and Buchberger (2006) examined the influence of transients and showed that the operation of a single water appliance inside a home is almost imperceptible in water mains and larger distribution network pipes and thus the sum of all residential demands of a single home can be used to define demands in a hydraulic model. They also showed that the (instantaneous) demand pulses deform in their path from the demand point to the upstream pipes. Thus, the assumption that the instantaneous rate of flow in a pipe is the sum of the concurrent downstream demands is a convenient approximation and one that is likely to be acceptable in most applications. McInnis and Karney (1995) calculated transients in a complex model from several pressure events using different models of demand aggregation. The model results could be improved (compared to available field data) by artificially damping the residual pressure waves and by increasing instantaneous orifice demands. This means that in transient models insight into demands is very important. Skeletonization also has an impact on hydraulic transient models (Jung et al. 2007), especially in modelling the periphery of the distribution network (as opposed to the larger diameter pipes or transport network).

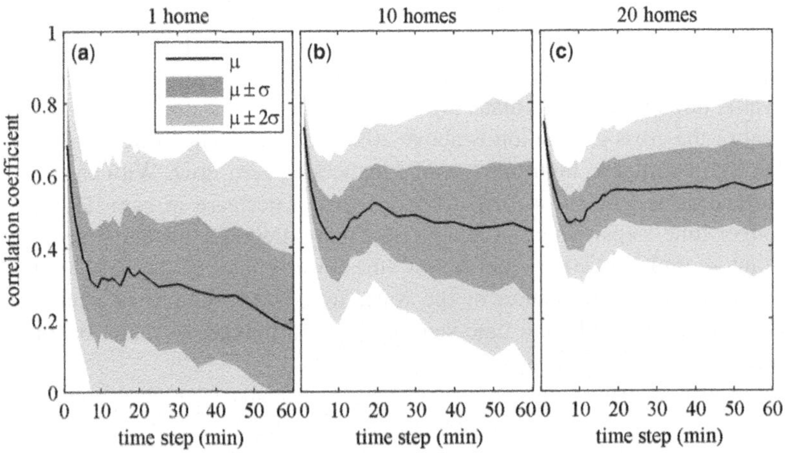

Figure 2.4 Mean (μ) and variation (σ) of lag-1 auto-correlation coefficient of measured patterns (Milford, OH; Buchberger et al. 2003) on different temporal and spatial scales; a) 1, b) 10 and c) 20 homes

The flow variance and scale of fluctuation, the probability of stagnation and the flow regime (laminar or turbulent flow) are affected by the time scale that is used in a water quality model (Li 2006; McKenna et al. 2003). Figure 2.5 shows for some typical (Dutch) flow patterns at different temporal scales and spatial scales (i.e. different number of downstream homes with appropriate pipe diameter) what the probability of stagnation, probability of laminar flow (Re < 2000) and probability of turbulent flow (Re > 4000) are. Above ca. 50 homes the time step has little effect on the probability of stagnation, laminar and turbulent flow. A small time step (<1 min) is mainly of interest in the end of the pipe system. Figure 2.6 shows, for the same demand patterns, the 95 to 100 percentile of the Reynolds number for different spatial and temporal scales. It makes clear that to determine the maximum flow a 1-minute time scale is necessary when demands from less than 200 homes are considered; if more than 200 homes are involved, a time scale of 5 minutes suffices.

Initial network simulations (1990s era) tended to use skeletonised distribution systems with a "top-down" demand allocation, a one-hour time step, and an advection-reaction (AR) water quality model (Rossman et al. (1994) is the classic example). This type of model can be calibrated with pressure measurements. Dispersion can be neglected where turbulent flows dominate (Li 2006). More recent analyses (since 2000) attempt finer resolution simulations using all-pipe networks with "bottom-up" demand allocation, a five-minute to one-hour time step, and an Advection-Dispersion-Reaction (ADR) water quality model (Li 2006; Tzatchkov et al. 2002). Here, top-down demand allocation means that the measured demand multiplier pattern of the pumping station is allocated to the demand nodes with a correction factor to account for total demand on that node, thereby applying strongly correlated demand patterns. A bottom-up demand allocation means that the demands are modelled per individual home and, subsequently, the individual demand patterns are summed to obtain the demand patterns at the demand nodes.

Using such a bottom-up approach, Buchberger and Li (2009) studied the effect of spatial correlation in the EPANET example 2 network, with demand nodes representing 1 to 140 homes and pipe diameters of 200 to 300 mm. They concluded that node-base demands are much more important than the spatial cross-correlation of demand patterns when considering the average tracer mass reaching the nodes. In the same network, Yang and Boccelli (2009) showed that demand variability can have a significant impact on when a chemical tracer arrives at a node; this is especially the case at the periphery of the network. With a tracer test in a real network, Blokker et al. (Chapter 6) showed that travel times can be as much as 30% longer on one day compared to another. These studies show that variation of demand between days may be a much more important effect than variation over the day or variation between nodes.

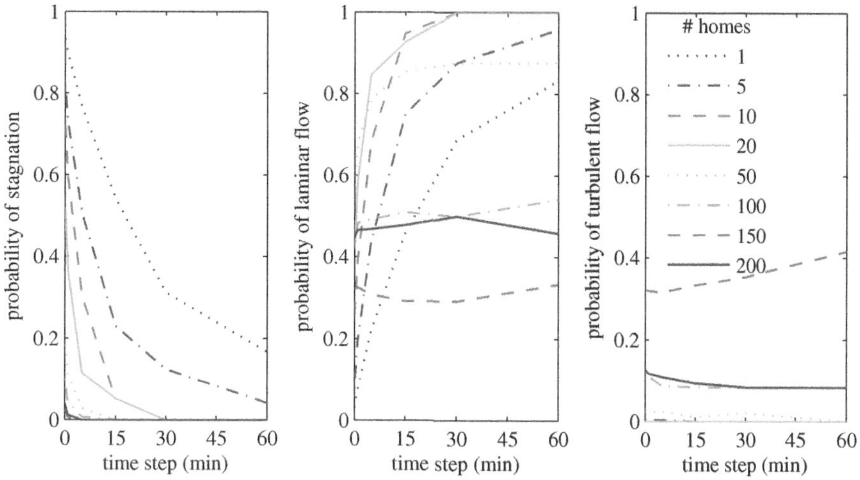

Figure 2.5 Probability of stagnation (Re = 0), laminar flow (Re < 2000) and turbulent flow (Re > 4000) for different time steps and number of homes (1, 5 homes: Ø59 mm; 10 homes: Ø100 mm; 20, 50, 100, 150 homes: Ø150 mm; 200 homes: Ø300 mm). The demand patterns to construct these graphs were simulated with SIMDEUM

Water quality modelling requires a detailed model of a distribution system. Demands must be known on a relatively small temporal (less than 5 minutes) and spatial (mains in a street) aggregation level and should be constructed by a bottom-up approach from demands of single homes. Since not every home can be modelled individually, a stochastic approach is required. In water quality modelling dispersion must be taken into account in an Advection-Dispersion-Reaction (ADR) model, especially if laminar-type flow conditions are expected to occur in the distribution mainlines (Li 2006; Tzatchkov et al. 2002).

2.5 DISCUSSION

Network water quality models on the distribution level may require fixture level or household level demands with no significant auto- and cross-correlation. This means that these models call for demand allocation via a bottom-up approach, i.e. allocating stochastic demand profiles with a small spatial aggregation level and appropriate short time scales.

There is currently no hydraulic network model that can properly work with instantaneous demands (i.e. on a per second basis) across an entire municipal network. Hence, even when nodal demands are known on a per second basis, they need to be integrated or averaged over a suitable time step before they can be used in a current network model. The best time step for hydraulic analysis will differ from the best time step for water quality analysis or human exposure analysis, and is related to the spatial aggregation level. When maximum flows are of importance (e.g. in sediment behaviour modelling) a suitable time step is one minute when less than 200 homes are considered; for larger spatial aggregation levels of five minutes would suffice, based on typical Dutch flow patterns. When the probability of stagnation is of importance (e.g. for modelling dissolved substances that are under the influence of dispersion and interact with the pipe wall), a suitable time step is one minute when less than 20 homes are considered; for more than 50 homes, a one-hour time step would suffice, based on data from the Milford field study. The question of the most suitable time step for network analysis needs further investigation. Also, the influence of using instantaneous demands on transient effects, water compressibility, pipe expansion, inertia effects, etc. in network models needs to be explored. Starting from very detailed network models with demands allocated per individual home and with time steps as short as one second, the effect of skeletonising and time averaging can be determined for different modelling purposes.

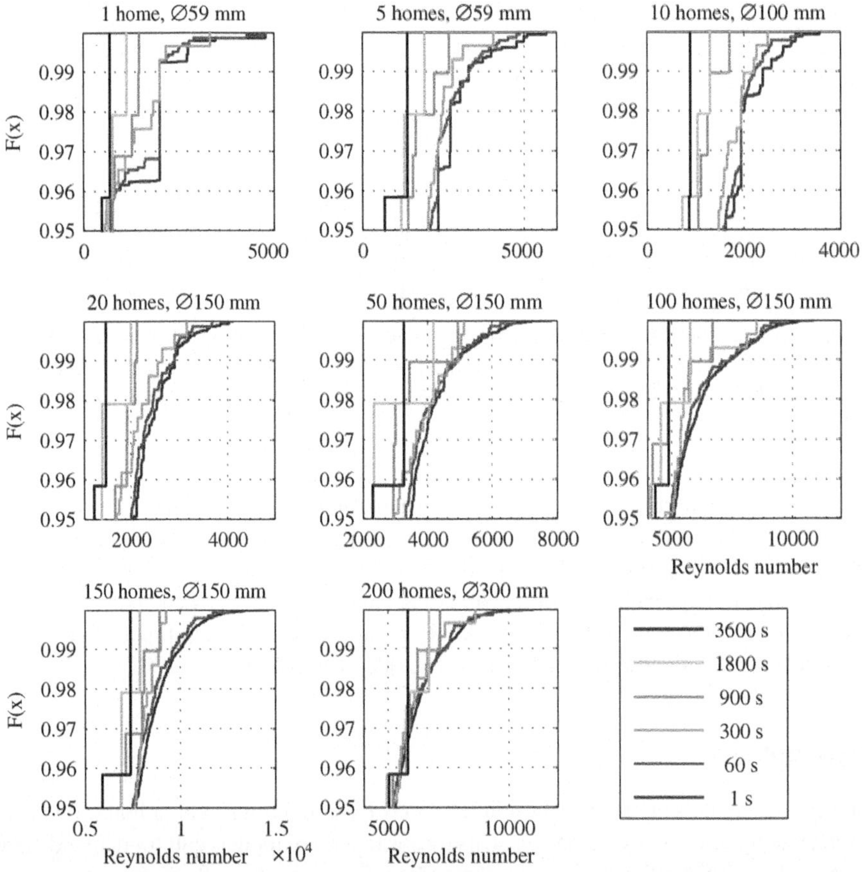

Figure 2.6 Maximum Reynolds number (95 to 100 percentile) for different time steps and number of homes (1, 5 homes: Ø59 mm; 10 homes: Ø100 mm; 20, 50, 100, 150 homes: Ø150 mm; 200: Ø300 mm). The demand patterns to construct these graphs were simulated with SIMDEUM

For a water quality network model, a stochastic demand model per (household) connection on a per minute or finer basis is needed. Today, two types of demand models are available that fulfil this requirement: the Poisson Rectangular Pulse model (Buchberger et al. 2003; Nilsson et al. 2005) and the end-use model SIMDEUM (Chapter 3; Chapter 5). Both the PRPSym and SIMDEUM demand models have been combined with hydraulic models in preliminary studies (Blokker et al. 2006; McKenna et al. 2005). So far, few water quality measurements were done to validate the model results. Li (2006) has applied PRPSym in combination with EPANET and an ADR-model to compare the model to measurements of fluoride and chlorine concentrations in a network. The ADR-model with the stochastic demand patterns gave good results with the conservative fluoride and reasonable results with decaying chlorine. In particular, predicted concentrations in the peripheral zone of the network showed much better agreement with field measurements for the water quality model with dispersion (ADR) than for the water quality model without dispersion (AR). Still, more network water quality models with stochastic demand should be tested with field data. This will reveal the shortcomings of the models and will indicate where improvement is to be gained. It will also provide more insight into the most suitable time step and spatial aggregation level for modelling.

Pressure measurements do not suffice for calibrating a network water quality model. Calibrating hydraulic models on pressure measurements typically means adjusting pipe roughness. This only affects pressures and not flows. Adjusting flows from pressure measurements is too inaccurate. An accuracy of 0.5 meters in two pressure measurements leads potentially to an uncertainty of 1 meter in head loss. On a total head loss of only 5 meters, this is a 20% imprecision in pressure and thus a 10% imprecision in flow. Calibrating a network water quality model requires flow or water quality measurements, e.g. through tracer studies (Jonkergouw et al. 2008).

With the use of stochastic demands in a network model, the question arises if a probabilistic approach towards network modelling is required and how to interpret network simulations. Nilsson et al. (2005) demonstrated that Monte Carlo techniques are a useful tool for simulating the dynamic performance of a municipal drinking water supply system, provided that a calibrated model of realistic network operations is available. A probabilistic approach in modelling and interpreting results is a significant departure from the prevailing practice and it can be used to complement rather than replace current modelling techniques.

2.6 CONCLUSIONS

Today, there is a growing interest in network water quality modelling. The water quality issues of interest relate to both particulate and dissolved substances, with the main interest in particles leading to discolouration, as well as affecting residual chlorine and contaminant propagation. There is a strong influence of flows and velocities on transport, mixing, production and decay of these substances in the network, which imposes a different approach to demand modelling. For transport systems the current hydraulic (AR) models suffice; for the more detailed distribution system, a network water quality model is needed that is based on short time scale demands and that considers the effect of dispersion (ADR) and transients. Demand models that provide trustworthy stochastic residential demands per individual home and on a one-second time scale are available.

The contribution of dispersion in network water quality modelling is significant. The contribution of transients in network water quality modelling still needs to be established. A hydraulics-based, or rather a stochastic demands-based, network water quality model needs to be developed and validated with field measurements. Such a model will be probabilistic in nature and will lead to a whole new way of assessing water quality in the DWDS.

Chapter 3

Simulating residential water demand with a stochastic end-use model*

ABSTRACT

A water demand end-use model was developed to predict water demand patterns with a small time scale (1 second) and small spatial scale (residence level). The end-use model is based on statistical information of users and end uses: census data such as the number of people per household and their ages; the frequency of use; duration and flow per water-use event; occurrence over the day for different end uses such as flushing the toilet, doing the laundry, washing hands, etc. With this approach, residential water demand patterns can be simulated.

The simulation results were compared to measured water demand patterns on attributes such as peak flow and daily total water use, as well as on the shape of the pattern and the frequency distribution of flows and accelerations in flow. The simulation results show a good correspondence to measured water demands.

The end-use model is based on independent statistical information and not on flow measurements. The input parameters are available before any information on annual or daily water use is available; the parameters are not fitted on flow measurements. Therefore, the model is transferable to diverse residential areas in different countries. The model can be applied in the design stage (pre-build), in scenario studies, and in water quality distribution network models.

3.1 INTRODUCTION

Water quality may change during transport and distribution. There is a requirement for more knowledge on the behaviour of both particulate and dissolved substances throughout drinking water distribution systems (Powell et al. 2004). The key element of a water quality model for a drinking water distribution system is a detailed hydraulic model (Slaats et al. 2003; Vreeburg 2010), which not only takes into account the maximum flows but also the flows on the preceding time steps (Powell et al. 2004; Slaats et al. 2003; Vreeburg and Boxall 2007). For that reason, knowledge of water demands is essential. For a water quality network model of drinking water distribution systems, a hydraulic model with an accurate probability of turbulent, laminar and stagnant flow is needed, and thus a detailed stochastic water demand model per (household) connection on a per second or per minute basis is required (Chapter 2).

---

*Reprinted with adaptations from Blokker, E. J. M., Vreeburg, J. H. G., and van Dijk, J. C. (2010). "Simulating residential water demand with a stochastic end-use model." *Journal of Water Resources Planning and Management*, 136(1), 19-26, doi:10.1061/(ASCE) WR.1943-5452.0000002 with permission from the copyright holders, ASCE.

Buchberger and Wu (1995) have shown that residential water demand develops from rectangular pulses; the pulses are described by their arrival time over the day, their intensity (flow) and duration. The parameters and probability distributions to constitute a Poisson Rectangular Pulse (PRP) model are derived from measurements (Buchberger and Wells 1996). The PRP model was applied in the USA (Buchberger et al. 2003), Italy (Guercio et al. 2001), Spain (García et al. 2004) and Mexico (Alcocer-Yamanaka et al. 2006). Different probability distributions for intensity and duration were found for different data sets, such as lognormal, exponential and Weibull distributions. Alvisi et al. (2003) use a model analogous to the PRP model based on a Neyman-Scott stochastic process (NSRP model), for which the parameters are also found by analysing measurements.

Obtaining the PRP parameters requires many (expensive) flow measurements. The parameters for Milford, Ohio (Buchberger et al. 2003), for example, were obtained from 30 days of measurements from 21 homes on a per second basis. The PRP model has only a few parameters and, therefore, it is a relatively simple model. The retrieved PRP parameters led to mainly short pulses of 1 minute or less. This means that showering, for example (circa 5 to 15 minutes), is almost never simulated as a single coherent event. Another issue is that it is difficult to determine how well the simulation performs compared to the actual measurements, since the simulation parameters were derived from the same or similar measurements. Also, it is difficult to correlate the parameters retrieved from these measurements with such data as population size, age and installed water-using appliances. As a consequence, the parameters for the PRP model are not easily transferable to other networks. The PRP model is a descriptive model, rather than a predictive one. The PRP model, thus, has a lot of potential to provide insight into some basic elements of water use, such as peak demands (Buchberger et al. 2008), travel times (Buchberger et al. 2003) and cross-correlation (Li et al. 2007).

Within the KWR Watercycle Research Institute, an approach based on end uses was developed to avoid large measurement campaigns. Within this approach each end use is simulated as a rectangular pulse from end-use specific probability distribution functions for the intensity, duration and frequency of use, and a given probability of use over the day (related to the residents' activities). Changes in appliances and water-use behaviour lead to different water demand patterns. An end-use model, therefore, acts as a predictive model and can be utilised in the design stage and in existing networks where no household water meters are installed.

3.2 METHODS AND MATERIALS – STATISTICAL ANALYSIS

The collected data were fitted onto several probability distribution functions; Table 3.1 describes the distributions that were used. The goodness-of-fit was determined with three parameters, viz. two measures for the error between empirical and modelled distributions (Mean Error, ME; and Root Mean Square Error, RMSE) and a measure for the similarity in shape between empirical and modelled distributions (coefficient of determination, R^2). The coefficient of determination is the proportion of variability in a data set that is accounted for by the statistical model. In case R^2 equals 1, the variation is completely explained by the model.

$$ME = \frac{1}{N}\sum_{i=1}^{N} y_i - x_i \tag{3.1}$$

$$RMSE = \sqrt{\frac{1}{N-1}\sum_{i=1}^{N} (y_i - x_i)^2} \tag{3.2}$$

$$R^2 = 1 - \sum_{i=1}^{N} (y_i - x_i)^2 \Big/ \sum_{i=1}^{N} (x_i - \bar{x})^2 \tag{3.3}$$

Here, x_i is the observed data, y_i is the model data and $\bar{x}$ is the mean of x_i. The ME and RMSE are expressed as absolute values and in percentages through comparison to the mean value of the measurement.

Table 3.1 Probability distribution functions and their mean and variance

Distribution	Probability Distribution Function	Mean	Variance
Normal	$y = f(x\|\mu, \sigma) = \frac{1}{\sigma\sqrt{2\pi}} e^{\frac{-(x-\mu)}{2\sigma^2}}$	μ	σ^2
Lognormal	$y = f(x\|M, S) = \frac{1}{S\sqrt{2\pi}} e^{\frac{-(\ln x - M)}{2S^2}}$	$e^{(M+S^2/2)}$	$e^{(2M+S^2)}\left(e^{S^2} - 1\right)$
χ^2	$y = f(x\|\nu) = \frac{x^{(\nu-2)/2} e^{-x/2}}{2^{\nu/2}\Gamma(\nu/2)}$	ν	2ν
Poisson	$y = f(x\|\lambda) = \frac{\lambda^x}{x!} e^{-\lambda} I(0, 1, \ldots)^x$	λ	λ
Negative Binomial	$y = f(x\|r, p) = \binom{r+x-1}{x} p^r (1-p)^x I(0, 1, \ldots)^x$	$\frac{r(1-p)}{p}$	$\frac{r(1-p)}{p^2}$

3.3 METHODS AND MATERIALS – MODEL DEVELOPMENT

3.3.1 Basic model

Buchberger and Wells (1996) have shown that the water demand pattern can be described by a Poisson Rectangular Pulse model. The end-use model is based on the same principle of rectangular pulses. The probability distributions describing the arrival time over the day, intensity (flow) and duration of the pulses in the end-use model are specified per end use. Furthermore, the parameters of the probability distributions are not retrieved from flow measurements but from statistical data from surveys. The end-use model produces a water demand pattern which can be described by the following equations:

$$Q = \sum_{k=1}^{M} \sum_{j=1}^{N} \sum_{i=1}^{F_{jk}} B(I_{ijk}, D_{ijk}, \tau_{ijk}) \tag{3.4}$$

$$B(I_{ijk}, D_{ijk}, \tau_{ijk}) = \begin{cases} I_{ijk} & \tau_{ijk} < T < \tau_{ijk} + D_{ijk} \\ 0 & elsewhere \end{cases} \tag{3.5}$$

Here, i, j and k are indices; k counts all end uses from 1 to M, j counts all users from 1 to N, i counts all busy times per end use from 1 to F_{jk} (the frequency of use for user j and end use k). D is the pulse duration (in seconds), I is the pulse intensity (flow in L/s) and τ is the time at which the tap is opened. Thus, D_{ijk} is the duration for end-use type k for user j and occurrence i. $B(I, D, \tau)$ is a block function, which equals I at time τ to $\tau + D$ and 0 during the rest of the day. The summation is done for all (M) available end uses, all (N) users and per end use for the frequency of use (F); this leads to the total water demand pattern Q (L/s) over the day. Figure 3.1 shows an example for 5 types of end uses and 2 users. All parameters are described by probability distribution functions which are unique for each end use.

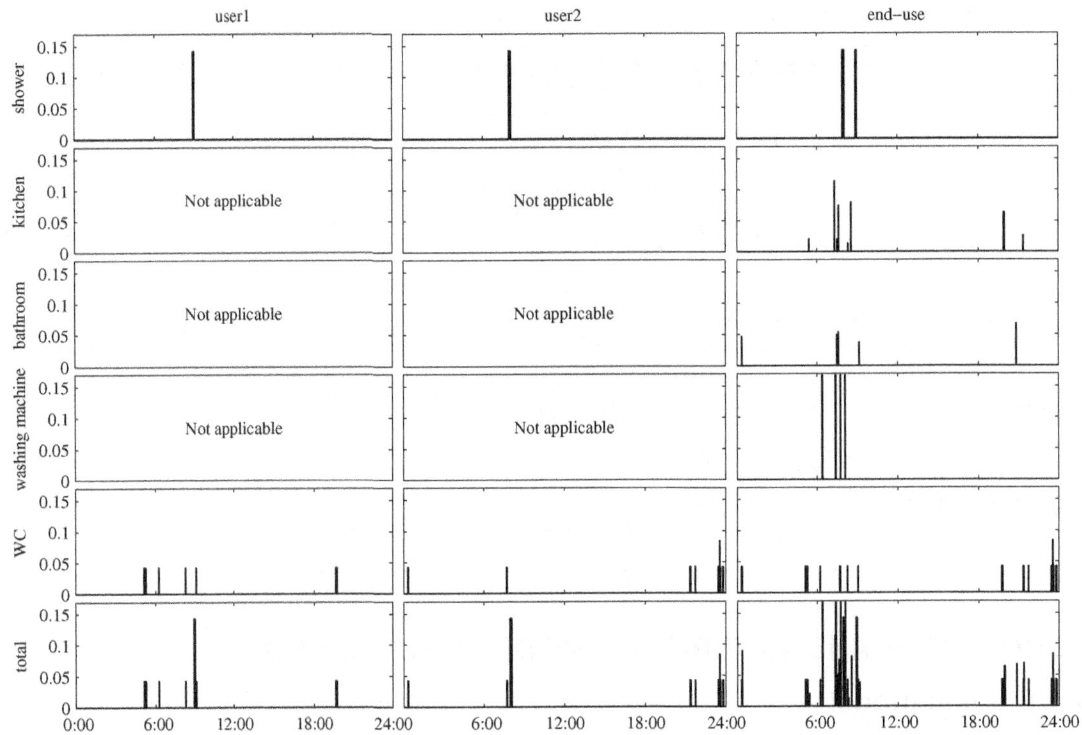

Figure 3.1 Example of summing water use for user $j = 1$ to 2 and end-use type $k = 1$ to 5 (viz. shower, kitchen tap, bathroom sink, washing machine, and WC)

3.3.2 Justification of input sources

Input sources for the end-use model are several surveys that were conducted independently of this research. The surveys on residential water use and time-budget are publicly available and are repeated frequently. Therefore, they are reliable sources of information. For the input of the end-use model, surveys of particular years were analysed in detail.

The Dutch water companies have a survey on residential water use carried out every three years since 1992. The particular survey that was analysed in depth was taken in 2001from about 3200 respondents (Foekema and Engelsma, 2001). They answered some general questions on their household (number of people, age and gender, and ownership of appliances) and fixtures (e.g., the flow during normal operation, measured with a stopwatch and a bucket with a specified volume). During one week they filled in a 'diary' on their water use at home. People were asked to write down how often they took a shower, flushed the toilet, opened the kitchen tap (and for what purpose), washed at the sink, took a bath, did the dishes (manually and with the dishwasher) and washed clothes. The data can be differentiated per age group and per household size. The survey was also done in 2004 (Kanne 2005a) and 2007 (Foekema et al. 2008). The 2007 data were not analysed statistically in this paper; if significant differences between 2007 and 2001 data were found, those data have been taken into account in this paper.

The Netherlands Institute for Social Research/SCP conducts the five-year time-budget survey since 1975. General questions on age, occupation, house and household were asked. During one week people wrote down in a diary what their main activity was (from a list of predefined activities) at every quarter of an hour of the day and whether they were at home or elsewhere. Unfortunately, these activities are not directly linked to water use, except for manually washing the dishes. Assuming a strong relationship between time of water use and times of getting out of bed, leaving the house, returning home and going to bed, this survey gives information on when water was possibly used. The time-budget survey that was used for the end-use model was the one from 1995 (SCP 1995). The 1995 survey was chosen because it had a large number of respondents: 3227 individuals in 1995, 1813 in 2000 and 2200 in 2005. Also, the surveys after 1995 gave less information on dishwashing as more and more households purchased a dishwashing machine.

3.3.3 The end uses (*k*)

Eight main types of end uses are defined at the fixture or appliance level, viz. water closet (WC), shower, washing machine, dishwasher, kitchen tap, bathroom tap, bathtub and outside tap. In the Netherlands, most of these main types have a penetration rate (number of households that possess a specific appliance) of 100% (Table 3.2). Dishwashers and bathtubs have a penetration rate that positively relates to the household size; the penetration rate of the bathtub also positively relates to a wealth class (defined by education level and income) and thus it relates to the type of house (price). The penetration rate of dishwashers is increasing (45% of about 3200 surveyed households in 2001, 54% in 2007) and the penetration rate of the bathtub is decreasing (Foekema and Engelsma 2001; Foekema et al. 2008).

For each type of end use there are different subtypes which constitute the end use (Table 3.2). For a WC, this can be an old-fashioned toilet with a large cistern of 9 to 12 litres or a new one of only 6 litres with a water-saving option of flushing only 3 litres. The subtype of a shower is determined by the combination of water heater and shower head, as they together determine the intensity (I). The subtypes of washing machines and dishwashers are different brands and types of machines with a specific water inlet pattern. A front-load washing machine (as mainly used in Europe) has a different inlet pattern and a considerably smaller total water use than a top-load washing machine (as mainly used in the USA). For the kitchen tap, a subtype is defined by its use, e.g., doing the dishes, consumption (water for meals and tea or coffee), washing hands and other uses. The types of end use are related to the frequency of use (F) and time of day (τ); the subtypes are related to intensity (I) and duration (D) of the water use event. A person flushes the toilet a number of times per day, regardless the type of toilet; the duration and flow at which the cistern is filled does depend on the type of toilet.

In Table 3.2, the penetration rate of the subtypes is the number of households that possess the subtype (e.g., type of toilet) or it is the distribution of use per subtype (e.g., for kitchen tap).

3.3.4 Users (*j*)

Users are key in an end-use model because they induce the end uses. Users are divided into groups based on household size, age, gender and occupation. Since the frequency of use is given per person, the number of people per household is of importance. A residential water-use survey in the Netherlands (Foekema and Engelsma 2001) showed several relationships: between age and frequency of use (e.g., older people flush the toilet more often and young children more often take a bath), between age and shower duration (teenagers take the longest showers), and between penetration rate and household size (e.g., possession of a dishwasher and a bathtub correlate positively with household size). It is possible to take these relationships into account and use an age-dependent frequency of use (F_{jk}) and duration (D_{ijk}). The time of water use (τ_{ijk}) is strongly related to the users' diurnal patterns (presence at home and sleep-wake rhythms). Consequently, the ages and occupations of users are important in the model as children (elementary school) and teens (secondary school or college) have different diurnal patterns from adults with jobs away from home or from senior citizens.

Table 3.2 Penetration rate or occurrence per end use in the Netherlands

End-use type	Penetration rate (%) in households	End-use subtype	Penetration rate (%) within end use
Bathtub	36	120 litres	100
Bathroom tap	100	Washing and shaving	33
		Brushing teeth	67
Dishwasher	45	Different brands and types	100
Kitchen tap	100	Consumption (drinking water, water for coffee and tea, water for cooking)	37.5
		Dishes and cleaning	25
		Washing hands	25
		Other (e.g., watering plants)	12.5
Outside tap	58	Garden	75
		Other	25
Shower	100	Kitchen geyser w/o water-saving shower head	8.3
		Kitchen geyser w/ water-saving shower head	8.3
		Bathroom geyser w/o water-saving shower head	8.3
		Bathroom geyser w/ water-saving shower head	8.3
		Combi-boiler w/o water-saving shower head	25.0
		Combi-boiler geyser w/ water-saving shower head	25.0
		Mini-boiler w/o water-saving shower head	8.3
		Mini-boiler geyser w/ water-saving shower head	8.3
Washing machine	98	Different brands and types	100
WC	100	High cistern (9 L)	11.1
		Low cistern (9 L)	22.2
		Low cistern (9 L) with water-saving option*	22.2
		Low cistern, new (6 L)	11.1
		Low cistern, new (6 L), with water saving option*	33.3

**When a water-saving option is available, it is applied in 80% of the flushings.*

Statistics Netherlands (CBS 2007) gives information on the number of households per city (or even per district) and the number of people, and their ages, within a household. Three major household types are discerned, viz. one-person households, two or more people without children (98.6% of these households consist of only two people; for the end-use model, 100% of this household type is assumed to consist of two people), and two or more people with children (with an average occupancy of 3.75 people). The average household size in the Netherlands is 2.3 people. For every household type, the number of people, the fraction of men and women, and the division over the different age groups is given (Table 3.3).

Table 3.3 Household statistics in the Netherlands as applied in the end-use model

		One person households	Two person households	Families with children
Number of people per household		1	2	3.75 (on average)
Number of households (%)		34	30	36
Gender division: Male / Female (%)		46 / 54	50 / 50	50 / 50
Age division (%)	Children (0–12 years old)	0	0	25
	Teens (13–18 years old)	0	0	16.5
	Adults (19–64 years old)	70	70	58.5
	Subdivision: % of adults with out-of-home job	Male: 67.5 Female: 52.4	Both persons: 49 Only male: 26 Only female: 6 Neither person: 18	Both parents: 39 Only father: 52 Only mother: 3 Neither parent: 5
Seniors (>65 years old)		30	30	0

3.3.5 The frequency of use (*F*)

For each end use, the frequency was retrieved from the residential water-use survey (Foekema and Engelsma 2001). The frequency is the number of uses per person per day; for the kitchen tap, however, it is the number of uses per household per day. The kitchen tap is often used for family purposes, such as cleaning, cooking and doing the dishes. Therefore, the frequency of use for the kitchen tap is less strongly related to individual users.

The frequency of use is described by a discrete statistical distribution, preferably the Poisson distribution because it has only one parameter that is easy to determine (i.e., the average λ). For six of the end uses, the frequency fits a Poisson distribution. For the kitchen tap, this is not the case because the high number of uses comes with a sample variance that exceeds the sample mean. Instead, a negative binomial distribution (an alternative to the Poisson distribution, where the sample variance is an explicit input parameter) is fitted on the data. Figure 3.2 shows the reported frequency of use of the kitchen tap and WC and fitted probability distributions. For the shower a binomial distribution is used, which means the number of showers is either zero or one. Table 3.4 summarises for all end uses how *F* can best be described.

3.3.6 The pulse intensity (*I*)

For each end use, the flow (in L/s) was determined (Table 3.4), partly from the water-use survey (Foekema and Engelsma 2001) and partly from technical information on water-using appliances as collected from installation guides (Vogelaar and Blokker 2004). For the shower, the intensity depends on the type of shower head as well as on the hot water supply. For the kitchen and bathroom taps, the maximum flow is determined by the pressure and the internal resistance of the indoor plumbing. The average flow during normal operation was equal to half the maximum flow from the technical data. Faucets are not always operated at the same flow. Therefore, a uniform distribution between 0 and the maximum flow is assumed. The flow was not measured for all end uses, so this could not be verified.

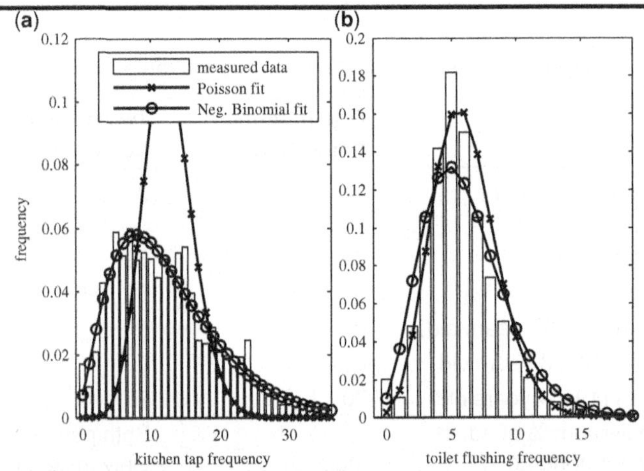

	Kitchen tap frequency		WC frequency	
	Poisson	Negative binomial	Poisson	Negative binomial
parameters	λ = 12.57	r = 3.00, p = 0.19	λ = 6.04	r = 8.77, p = 0.59
ME	-6.9e–15	-5.8e–5	8.8e–18	2.9e–18
RMSE	0.0252	0.0049	0.0077	0.0089
R2	51%	94%	96%	93%

Figure 3.2 Reported frequency of use plus fitted Poisson and Negative binomial distribution of a) the kitchen tap per household per day and b) WC flushes per person per day

3.3.7 The pulse duration (*D*)

For each end use, the duration was determined (Table 3.4), partly from the water-use survey (Foekema and Engelsma 2001) and partly from technical information on water-using appliances. The duration of taking a shower is typically determined by the user, while the duration of filling the toilet bowl is determined by the volume of the bowl and the inlet flow which is related to the installation. The duration of taking water from a bathroom or kitchen tap, which is determined by the user, can be described using a lognormal distribution function.

For the kitchen tap, a lognormal distribution fits the data, with the variance equal to 130% of the mean value. For the durations of the bathroom tap and outside tap, the averages are known, but the standard deviations are not. A lognormal distribution with a variance equal to 130% of the mean is assumed. For the duration of showering, a lognormal distribution with the variance equal to 50% of the mean value or a χ^2-distribution can be used (Figure 3.3). However, the χ^2-distribution is only valid when the duration is expressed in minutes because this frequency distribution is not linearly scalable.

Table 3.4 Frequency, duration and intensity for several types and subtypes of end uses in the Netherlands (Blokker 2006), average (μ) and probability distribution function (pdf)

End-use type / subtype		Frequency (day^{-1})		Duration		Intensity (L/s)	
		μ	pdf	μ	pdf	μ	pdf
Bathtub	120 litres	0.044	Poisson	10 min	N.A. (fixed)	0.200	N.A. (fixed)
Bathroom tap	Washing and shaving	4.1	Poisson	40 s	Log-normal	0.042	Uniform
	Brushing teeth			15 s			
Dish washer	Brand and type	0.3	Poisson	Specific dishwashing pattern (4 cycles of water entering, total 84 seconds, 0.167 L/sec = 14 L)			
Kitchen tap	Consumption	12.6*	Negative binomial (r = 3, p = 0.192)	16 s	Log-normal	0.083	Uniform
	Doing dishes			48 s		0.125	
	Washing hands			15 s		0.083	
	Other			37 s		0.083	
Outside tap	Garden	0.44	Poisson	300 s	Log-normal	0.1	Uniform
	Other			15 s			
Shower	Normal	0.7	Binomial	8.5 min†	χ^2	0.142‡	N.A. (fixed)
	Water saving type					0.123	
Washing machine	Brand and type	0.3	Poisson	Specific washing pattern (4 cycles of water entering, total 5 minutes, 0.167 L/sec = 50 L)			
WC	6-litre cistern	6.0	Poisson	2.4 min§	N.A. (fixed)	0.042	N.A. (fixed)
	9-litre cistern			3.6 min			

**the frequency for the kitchen tap is per household per day.*
†shower duration has an age dependency; children and teens take longer showers.
‡the shower intensity depends on the type of water heater.
§with a water saving option the duration is reduced to 50% of the original value.

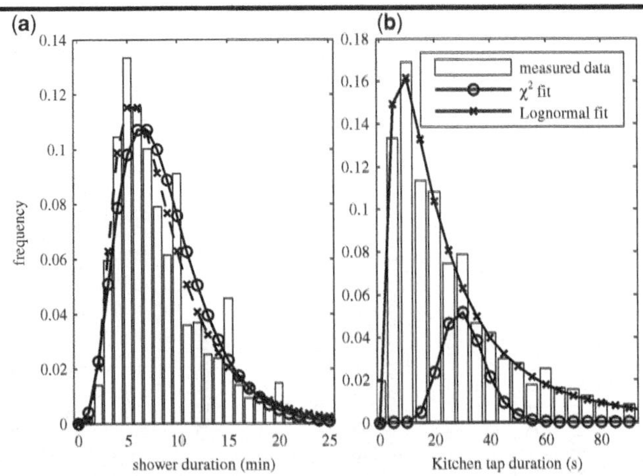

	Shower duration (min)		Kitchen tap duration (s)	
	χ^2	Lognormal	χ^2	Lognormal
parameters	ν = 8.5	M = 2.0, S = 0.5	ν = 30.6	M = 3.0, S = 1.0
ME	1.0e–6	–1.9e–5	–0.015	–2.1e–4
RMSE	0.0116	0.0082	0.0372	0.0056
R^2	89%	95%	n.a.	98%

Figure 3.3 Duration plus fitted χ^2 and Lognormal fit of a) shower (1 min bins) for adults (18–64 years old) and b) opening kitchen tap (5 s bins) for washing hands

3.3.8 The diurnal pattern, time of water use (τ)

There is no survey available that addresses for all end uses the time of the day when a water-use event is likely to take place. For the washing machine and dishwasher, a question was included in the residential water-use survey (Foekema and Engelsma 2001) about when these appliances were mainly used: morning, afternoon, evening or night. For doing the dishes by hand, a much more accurate pattern can be constructed based on the time-budget survey (SCP 1995). This pattern correlates to the pattern of preparing a meal and dining (Figure 3.4). For these activities a (rough) pattern can be constructed to estimate τ.

For the other types of end uses, it is assumed that the start of water use is strongly related to whether people are at home or not and if they are asleep, getting up or preparing for bed (Figure 3.4). From the time-budget survey (SCP 1995), the duration of sleep and being away and the time of getting up and leaving home (in the morning) can be fitted to a normal distribution (Figure 3.5, Table 3.5). For all four parameters, a chi-square test failed to confirm on a 5% significance level that the data are normally distributed. As an alternative to using the data from the time-budget survey directly, the end-use model uses the normal probability distributions to draw random numbers because it simplifies the model, and the difference between the two approaches is small (Figure 3.5). A diurnal pattern for the probability of water use is thus constructed from the diurnal pattern of users, i.e., their sleep-wake rhythm and their being at home. For weekdays, this pattern is strongly related to people's ages and occupations;

for weekend days, this correlation is not very strong and the same parameters are assumed for all age groups.

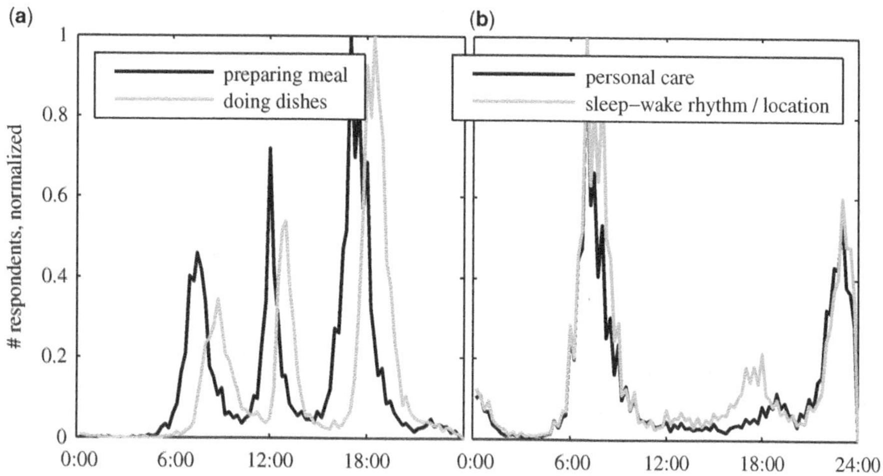

Figure 3.4 Diurnal pattern for a) preparing a meal and doing the dishes, b) personal care over the day compared to getting up, leaving the house, returning home and going to bed

A diurnal pattern for a user can be constructed with random picks from the normal distribution. This pattern is converted to a probability distribution function (PDF) of water-use events in the following way:

- During the sleeping hours, the total volume of water use is estimated at 1.5% of the total daily demand, based on the analysis of Dutch water use measurements between 1:00 AM and 5:00 AM (Blokker 2006).
- During absence, the probability of (the start of) water use is set to zero.
- During the half hour after getting up and returning home and the half hour before leaving home and going to bed, peak hours are assumed and the fraction of the total demand occurring in this period is estimated at 65%. The 65% is based on the assumption that, during the peak hours, 100% of washing (use of shower, bathtub and bathroom tap) takes place (which is 50% of the average Dutch water use), 50% of flushing the toilet and 10% of all other water uses take place.
- The rest of the water use then corresponds to 33.5% of the daily water use.

The cumulative distribution function (CDF) is the cumulative sum of the PDF and is scaled to be between zero and one. Since the times of waking up, leaving the home, returning home and going to bed vary (Table 3.5) the probability of water-use events varies as well. A random time of water use (τ) can be retrieved from the CDF. An example is shown in Figure 3.6 where a random number (from a uniform distribution between 0 and 1) equal to 0.45 means that τ for the water-use event is 18:15.

For *Bath*, *Bathroom tap*, *Outside tap*, *Shower* and *WC*, the CDF over the day is constructed from Table 3.5 as described above. For *Kitchen tap*, *Dishwasher* and *Washing machine*, the CDF is constructed by multiplying the pattern from Table 3.5 with the respective patterns for dishwashing by hand or machine and washing machine use, leading to Figure 3.7.

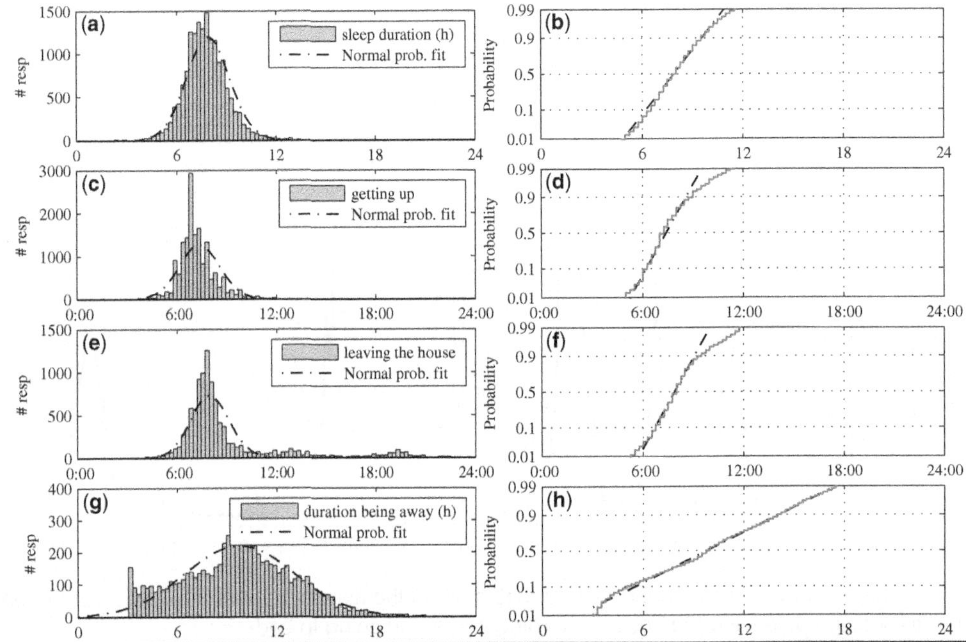

	Sleep duration	Time of getting up	Time of leaving the house	Duration of being away
parameters	μ = 8.0, σ = 1.3	μ = 7:25, σ = 1.3	μ = 8:00, σ = 1.2	μ = 9.8, σ = 3.3
ME	−1.9e–20	−1.1e–19	−1.5e–19	-5.8e-19
RMSE	0.004	0.014	0.009	0.004
R^2	96.5%	62.3%	82.8%	86.5%

Figure 3.5 Frequency distribution and cumulative frequency distribution (in normal probability plot) for (a, b) the duration of sleep, (c, d) time of getting up, (e, f) time of leaving the house and (g, h) duration of being away from home (for >3 hours) for all respondents of the 1995 survey

3.4 METHODS AND MATERIALS – THE SIMULATION

Once the statistical information on users and water outlets per house are put into the model, the simulation can be run. The simulation scheme is depicted in Figure 3.8.

First, a household is simulated: the occupancy and the age and gender (from Table 3.3) and diurnal patterns of the occupants (Table 3.5) are determined. The input statistics can be specific for certain types of houses. For example, small apartments hardly ever house more than two people; special apartments for elderly people will not be inhabited by families with children. Next, it is determined what end uses ($k = 1$ to 8) and subtypes are present in each simulated house (Table 3.2; for the human operated taps the subtype is determined per occurrence). Then, per end use k (in random order), for all users ($j = 1$ to N) in the house, the frequency of use (F_{jk}) is determined from the appropriate probability distribution function (often a Poisson distribution, Table 3.4). After that, for all occurrences ($i = 1$ to F_{jk}), the duration and intensity are determined from the appropriate probability distribution function (often a lognormal distribution for the duration and a uniform distribution or fixed value for the intensity, Table 3.4) and when during the day (τ) the water-use event occurs (from the probability function per person, Figure 3.6 and Figure 3.7). When for end use k, user j and occurrence i the time τ_{ijk} is established, the time $\tau_{ijk} - D_{ijk} < T < \tau_{ijk} + D_{ijk}$ is blocked for end use k, user $j+1$ to N and occurrence $i+1$ to F_{jk}. This ensures that there is no overlapping use by multiple users of one end use, e.g., when only a single shower is present, the use of it by one person cannot coincide with its use by a second person. However, user j could potentially draw water from multiple end uses (k and $k+1$) at the same time. This is not actively blocked, but appears to occur seldom in the simulation. The sum of all water demands is the water demand pattern of the home.

Table 3.5 Statistics for diurnal patterns in the Netherlands

		Weekday					Weekend day
		Child	**Teen**	**Adult with out-of-home job**	**Adult without out-of-home job**	**Senior**	
Time of getting up	μ	7:00	7:00	7:00	8:00	8:00	9:00
	σ	1:00	1:00	1:00	1:00	1:00	1:30
Time of leaving the house	μ	8:30	8:15	8:00	13:00	13:00	13:00
	σ	0:30	0:30	0:45	3:00	3:00	3:00
Duration of being away	μ	7.0 h	8.0 h	9.5 h	10.0 h	10.0 h	10.0 h
	σ	2.0 h	2.0 h	3.25 h	4.5 h	4.5 h	4.5 h
Duration of sleep	μ	10.0 h	9.0 h	7.0 h	8.0 h	8.0 h	9.0 h
	σ	1.0 h	1.0 h	1.0 h	1.0 h	1.0 h	1.5 h

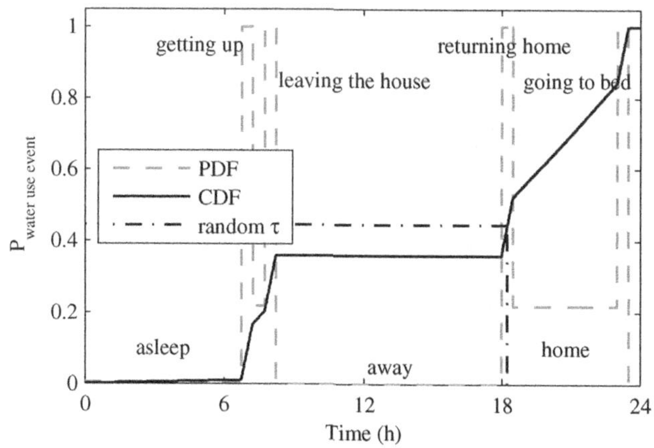

Figure 3.6 An example of drawing a random time (τ) from the CDF of water use for a working adult

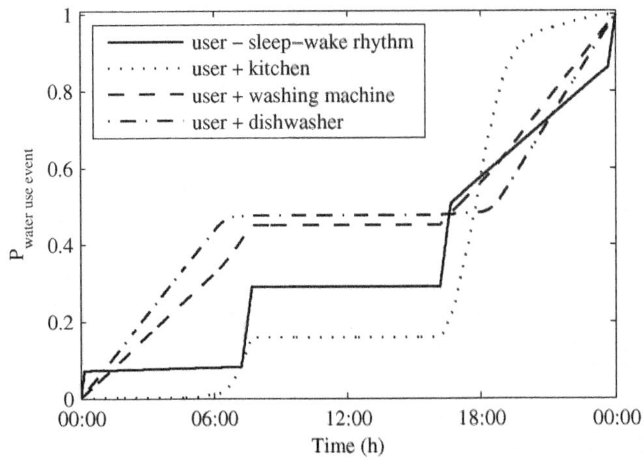

Figure 3.7 CDF of the probability of water use events for 1) end uses that depend on just the sleep–wake rhythm of a user (WC, shower, bathtub, bathroom tap, outside tap); 2) use of the kitchen tap; 3) use of the washing machine and 4) use of the dishwasher

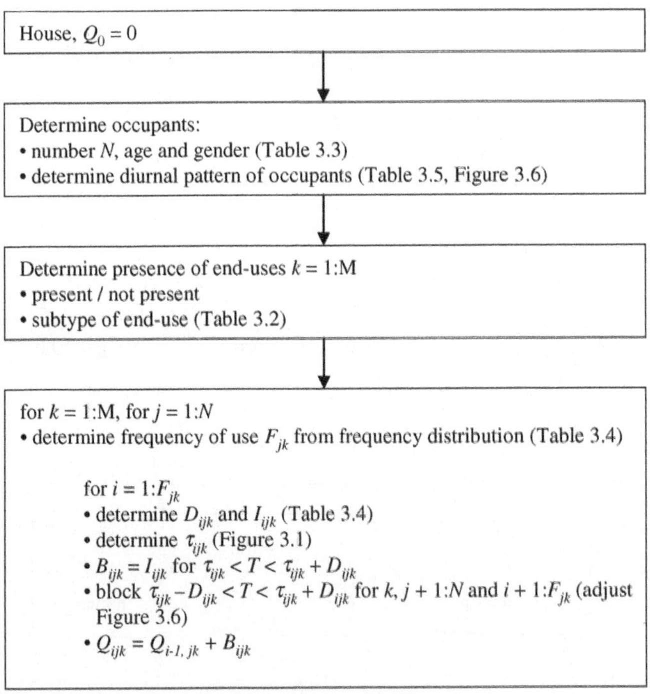

Figure 3.8 Simulation scheme

The output of one simulation run is one water demand pattern (Figure 3.9a). By doing more simulations, either multiple patterns (days) of a specific house are found or a water demand pattern of a street with several houses on it could be constructed (Figure 3.9b). Due to the stochastic nature of water demand patterns, this single simulated water demand pattern is only one possible outcome. With a Monte Carlo simulation, many simulation results are obtained and a statistical analysis can be made on the results. The end-use model was programmed in MATLAB® to run the simulations. The software model was named SIMDEUM: SIMulation of water Demand; an End-Use Model.

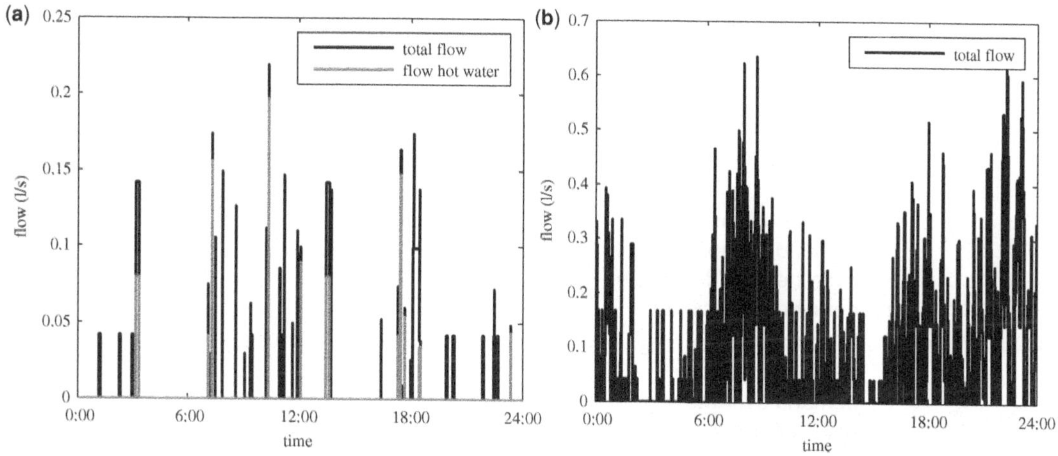

Figure 3.9 Simulation results at a time scale of 1 s a) demand pattern of a single house and b) demand pattern for a street with 20 houses

3.5 METHODS AND MATERIALS – MODEL VALIDATION PARAMETERS

To establish the performance of the water demand model, the results need to be compared to water demand measurements on several characteristic parameters. Since the output of the model is stochastic in nature, the frequency distribution of the characteristic parameters is used in the comparison. To quantify the validation, some statistic measures were used.

For characteristic parameters, García et al. (2004) suggest Q_{max} (the maximum flow on different time scales), V (the volume per day), the number of pulses per day and the number of clock hours with non-zero water use per day. For all these parameters, both the average and probability distributions are considered. Furthermore, the mean and variance of water demands over the day (Q_{day}) on different time scales are used. Q_{day} can also be converted to a dimensionless water demand multiplier pattern (C_d). The cumulative frequency distribution of water demands (Q_{CFD}) and the cumulative frequency distribution of the change in water demands (ΔQ_{CFD}) are introduced as extra parameters. These parameters, and the time and spatial scales that they are valid for, are summarised in Table 3.6.

Validation on different time scales (from 1 second to 1 hour) and different spatial scales (from an individual house to a street of 5–50 homes to an isolated section of 100–500 homes) needs to be considered. The reason is that water demand models are applied on different scales, not only in a network model, but also in design and scenario studies. A good match between measurement and simulation on a small time (or spatial) scale does not automatically lead to a good match on larger time (or spatial) scales.

Table 3.6 Overview of parameters used for validating simulation results. The temporal and spatial scales are independent variables

Parameter		Description	Temporal scale	Spatial scale
Q_{max}	(L/s)	Maximum flow per day	1 s, 1 min, 1 h	house, street, isolated section
V	(L)	Total water use per day	1 day	house, street, isolated section
n_{pulse}	(-)	Number of pulses per day	1 s	house
n_{hours}	(-)	Number of hours of water use / busy hours	1 h	house
Q_{day}	(L/s)	The water demand over the day	10 min, 1 h	street, isolated section
C_d	(-)	Dimensionless water demand multiplier pattern over the day (daily water demand coefficient)		
Q_{CFD}	(L/s)	Cumulative frequency distribution of water demand over the day (CFD of flows)	1 s, 1 min, 15 min	house, street
ΔQ_{CFD}	(L/s^2)	Cumulative frequency distribution of change in water demand over the day (CFD of flow accelerations).		

Not all parameters are suitable for all time scales and all spatial scales. The number of pulses per day is only of interest with small time scales; measurements on a five-minute basis will give no information about how many times a tap was opened within that time. To compare the actual patterns, Q_{day} is most suitable for larger time and spatial scales (about 10 minutes or longer and the spatial scale is at least the sum of 15 houses). For smaller time scales (1 second to circa 15 minutes) and smaller spatial scales (e.g., a single house), i.e. for demand patterns with little auto-correlation, Q_{CFD} and ΔQ_{CFD} are used. Note that the minimum applicable time scale also is strongly related to the accuracy of the available measurements.

To get an accurate statistical overview, the number of measurements that the simulations are compared against is important. The parameters Q_{day}, C_d, Q_{CFD} and ΔQ_{CFD} are based on all measurements over the day and, thus, an accurate overview is already acquired with a few measurement days. The other parameters are based on one number per day (e.g., maximum daily flow) and, therefore, at least 25 days' worth of measurements are required.

The parameters Q_{max}, V, n_{pulse}, n_{hours}, Q_{CFD} and ΔQ_{CFD} are presented as cumulative frequency distributions. The goodness-of-fit for the cumulative frequency distributions is determined with ME, RMSE and R^2 (Eqs. 3.1 to 3.3) evaluated at each integer percentile between 1 and 100 (1, 2, 3, … 100). Q_{day} is a pattern; the goodness-of-fit is expressed as ME, RMSE and R^2 of Q_{day}. In this paper the number of pulses per day is not used, due to the lack of detailed measurement data. The ME and RMSE are presented as absolute values and as percentages of the means of the measurements. For Q_{CFD} and ΔQ_{CFD} the relative values are not meaningful because the averages of the measurements are close to 0.

3.6 METHODS AND MATERIALS – MODEL VALIDATION

The results that are presented here are comparisons of available measurements and simulations that were done specifically for this purpose.

In 2004, Waternet (the water company of Amsterdam) logged the water use of 46 homes dispersed over the city; each home was measured for 7 days. A standard water meter with a nominal flow of 1.5 m^3/h and an external pulse was used. The logging frequency was one minute; the logging precision was 1 litre per pulse. This resulted in errors due to rounding. Therefore, the water demand patterns were aggregated to a five minute time scale. One of the homes had a 0.2 L/min leak; this house was not considered in the validation because SIMDEUM does not simulate leakage. The measurements of two other homes were excluded because they had extremely high water uses which casted doubt on the accuracy of the measurements. The measurements were not all done in the same week. Water demand patterns for the sum of the remaining 43 homes were constructed by adding the measured patterns of equal weekdays.

At the original 46 homes, a residential water-use survey was conducted (Kanne 2005b). This led to accurate knowledge of the number of people per home and specific information on water use at the measured homes. Compared to Table 3.2's data that looked at locations throughout the Netherlands, the 46 "measured" houses in Amsterdam had fewer water-saving options on toilets (48% instead of 56%), fewer water-saving shower heads (20% instead of 50%), different hot water devices (35% kitchen geyser, 53% combi- or mini-boilers and 12% collective hot water supply), and fewer outside taps (23% instead of 58%). Compared to Table 3.3, the household composition in Amsterdam is also different. There are more one-person households (55% instead of 34%), fewer two-person households (21% instead of 30%), fewer families with children (24% instead of 36%), and the average household size is smaller (1.8 instead of 2.3). The average age is younger, only 11% (instead of 19%) of the population is over 65 years old (CBS 2007).

Simulations were done using SIMDEUM, with the default input parameters plus the specific information on households and in-home installations. Only weekday patterns were considered; no specific weekend patterns were taken into account. The statistical results were retrieved from 1000 simulated patterns over 10 days at 100 different houses. The simulations were done on a 1 second time scale and time-averaged over 5 minutes.

3.7 RESULTS

Figure 3.10 and Figure 3.11 and Table 3.7 show that there is good agreement between simulations and measurements (ME and RMSE $< 30\%$ and $R^2 > 0.7$) of individual homes. The maximum flow on different time scales and volume per day are predicted well by the end-use model. The pattern is defined by the parameters n_{hours}, Q_{cdf} and ΔQ_{cdf}; these parameters are predicted well by the end-use model.

Figure 3.12 shows that Q_{day} is well predicted for the sum of 43 homes; the peaks in the morning coincide in time and maximum flow, the minimum values are also predicted well. The average water use in the simulation, however, is about 17% too low and the R^2 of the total pattern is low (Table 3.7) which relates mainly to afternoon consumption. The reason for the lower simulated average is probably that in Amsterdam more people are at home than the end-use model assumes; in Amsterdam people work fewer hours than the overall Dutch average (CBS 2007).

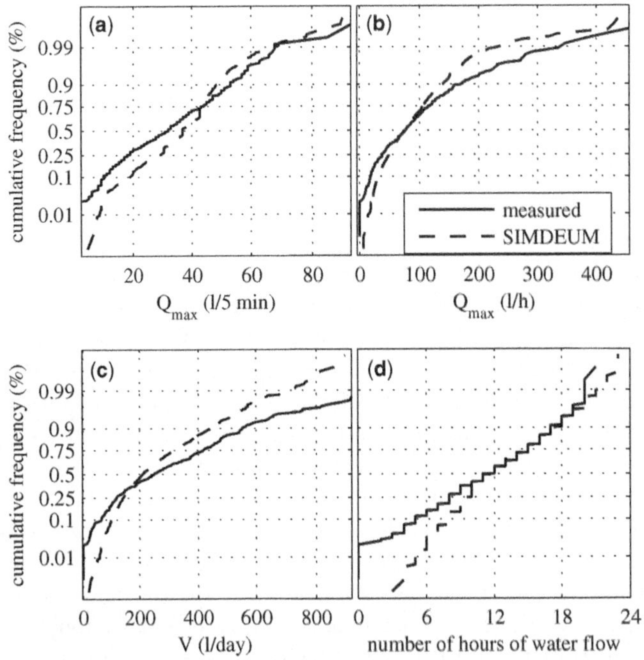

Figure 3.10 Comparing, at a time scale of 5 min and spatial scale of 1 home, measurements (43 houses, 7 days each) and simulations (100 houses, 10 weekdays at 1 s time scale time-averaged over 5 min); a) Q_{max} per 5 min, b) Q_{max} per h, c) V, d) n_{hours}

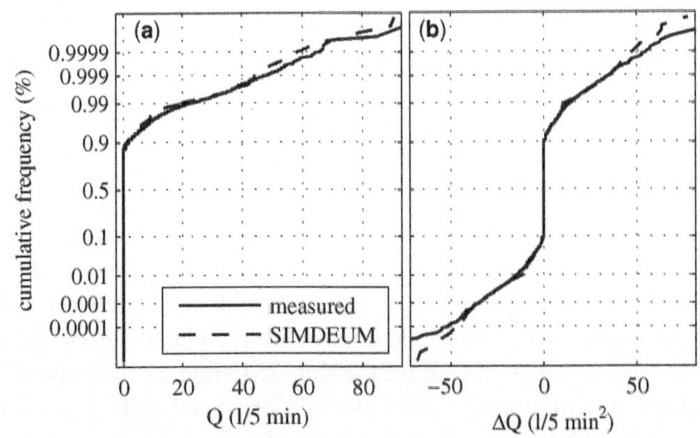

Figure 3.11 Comparing, at a time scale of 5 min and spatial scale of 1 home, measurements (43 houses, 7 days each) and simulations (100 houses, 10 weekdays at a 1 s time scale time-averaged over 5 min); a) Q_{CDF}, b) ΔQ_{CDF}

Table 3.7 Statistics of comparison between measured and simulated water demands. All values are statistically significant

Parameter	Temporal scale	Spatial scale	Graph	ME	RMSE	R^2
Q_{max} (L/5 min)	5 min	1 home	Figure 3.10a	4.3 (13.6%)	6.8 (21.3%)	0.84
Q_{max} (L/h)	1 h		Figure 3.10b	−5.4 (−6.0%)	22.9 (25.5%)	0.91
V (L/day)	1 day		Figure 3.10c	−46.0 (−15.5%)	92.4 (31.2%)	0.83
n_{hours}	1 h		Figure 3.10d	0.9 (7.6%)	1.53 (13.3%)	0.89
Q_{CDF} (L/5 min)	5 min		Figure 3.11a	−0.19 (N.A.)	0.64 (N.A.)	1.00
ΔQ_{CDF} (L/5 min^2)	5 min		Figure 3.11b	−0.11 (N.A.)	1.15 (N.A.)	0.98
Q_{day} (L/5 min)	5 min	43 homes	Figure 3.12	−7.6 (−17.4%)	23.9 (54.8%)	0.19

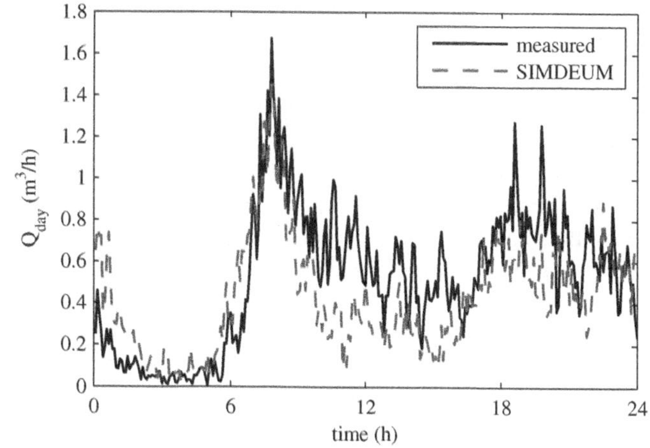

Figure 3.12 Comparing average Q_{day}, at a time scale of 5 min and spatial scale of 43 homes, of measurements (sum of 43 houses, 5 weekdays each) and simulations (sum of 43 houses, 10 weekdays at a 1 s time scale time-averaged over 5 min)

3.8 DISCUSSION

Building a water demand model from independent statistical information on users, end uses, frequency of use, intensity and duration per use, and time of water use leads to realistic water demand patterns that compare well to measured water demand patterns. The assumption of a strong correlation between time of water use and the users' sleep-wake rhythm and their being at home appears to be valid.

The end-use model could be improved or extended in several ways, which can be related to the parameters of Eq. 3.4:

- Other types of water use (k), such as leaks, could be incorporated. Dripping taps or other types of leakage could form a substantial part of the total water demand (Cobacho et al. 2004). The 2007 residential water-use survey (Foekema et al. 2008) showed that, in the Netherlands, about 5% of the households have dripping taps, 2% have leaking toilets and fewer than 1% have leaking pipes.

- The intensity (I) of most end uses is assumed to be deterministic or uniformly distributed. From measurements (e.g. Buchberger et al. 2003), a lognormal probability would be more likely.
- The duration (D) of the end uses is mainly assessed as a lognormal probability with the variance equal to 130% of the mean value. This is not necessarily true for all cases. Improving the model with better estimates of duration would require measurements of micro-components.
- The frequency of use (F) was determined for all end uses. The use at the kitchen tap showed that a Poisson distribution was not appropriate. Instead a Negative Binomial distribution was applied. The downside is, that this distribution requires two parameters that can not easily be determined from the average number of uses. For the shower, a binomial distribution was applied, because the Poisson distribution did not fit on the data of 2001 (Foekema and Engelsma 2001). However, the survey of 2007 (Foekema et al. 2008) showed that the shower frequency is increasing and a Poisson distribution will be better fit to describe this.
- Especially uncertain are the duration and frequency of use of outdoor water use. The water-use surveys have not been focussing on outdoor water use. The water companies, however, do see that the yearly peak demand does not occur the morning (as it is on normal days), but in the late afternoon on warm summer days. This is related to outdoor water use. In dry countries, outdoor water use is much more significant than in the Netherlands. With the expected climate change, outdoor water use may become more important in the Netherlands as well. It is recommended that the future water-use surveys will include more detailed questions on outdoor water use.
- The probability of a water-use event over the day (τ) is based on the assumption of a relation with the users' sleep-wake rhythm and 65% of the water use events occur during the peak hours. The occurrences of taking a shower during the day are not explicit in the model due to a lack of information. In the water-use survey of 2007 (Foekema et al. 2008) an extra question was included to determine when, over the day, the shower was being used. Most people indicated taking a shower after getting up in the morning (65%), 42% take a shower before going to bed and 17% take a shower at other times (e.g., after physical exercise). This specific information may improve the model.
- Demand could be a function of network pressure, i.e., with lower pressure I decreases and D may increase. It would be interesting to further investigate how this can be modelled.
- In the modelled households, the house mates' sleep–wake rhythm and presence at home are independent. In reality, family members tend to tune their diurnal patterns. For water demand this independency may not be an issue, since the simultaneous use of one appliance or fixture by more than one person is inhibited within SIMDEUM. For energy demand patterns it is more important if several people are at home together (and thus share heating and lighting) or that they are at home independently of each other (and thus use heating and lighting consecutively).

A thorough sensitivity analysis is required to determine how accurately parameters must be known. This will provide insight into the applicability of the end-use model and into what improvements will be most effective.

Another expansion of the model can be towards its application outside of the Netherlands. An end-use model can easily be constructed for various neighbourhoods in different countries, even in the design stage, provided the statistics are available. The Centre for Time Use Research (2007) has links to data from time-use surveys all over the world. Literature on water use, often aimed at coarse water demand management, is available for many countries (e.g. Jacobs and Haarhoff 2004; Memon et al. 2007; Vieira et al. 2007; White et al. 2004). With the information on averages and frequency distributions of the frequency of use, duration and intensity per end use from the Dutch data and some country specific information, input into the end-use model for other regions can be determined. For the Netherlands, 156 SIMDEUM parameters were determined. A minimum set of parameters, which means not taking the influence of age, gender and household size into account and not define end-use subtypes, is 36. Of these 36 parameters, some can be reused from the Dutch data.

The model is based on statistical information about in-home installations and residents. Therefore, the influence of an aging population on the total and peak water demand, of the decrease in household size, or of the replacement of old appliances with new ones can be easily determined. The end-use model thus enables quantitative scenario studies. Because no flow measurements are needed as an input, the model can be employed before a drinking water distribution system is built and can then be utilised in the design stage and in existing networks where no household water meters are installed.

With the end-use model, a realistic consumption estimation is available and new applications, therefore, come into sight. Water-demand allocation in a water distribution network model can be done without using measured water demand patterns. This allows for more detailed water quality modelling in the periphery of the drinking water distribution system.

3.9 CONCLUSIONS

A stochastic end-use model for the simulation of residential water demand was developed. The end-use model is based on independent statistical information of water-using appliances and (residential) users instead of water demand measurements. The frequency of water use is mainly determined by a Poisson distribution; a negative binomial distribution is applicable to the frequency of use for the kitchen tap. The intensity of water use depends on the type of end use and was described by a constant or a uniform probability distribution. The duration of water use is either determined by the user and can be described by a lognormal distribution, or by a water-using appliance and can then be described by a constant. The diurnal water use was estimated by using statistical information of users' activities, such as their time of going to bed and getting up, leaving the house and returning home. With limited input information, a pattern was predicted for a fraction of the costs involved in the conventional measuring approach.

First results show that the simulation results are in good agreement with measured water demand patterns. This is true for a range of time scales (from 1 minute to one hour) and a spatial scale of a single home and the sum of 43 homes.

Chapter 4

Simulating non-residential water demand with a stochastic end-use model*

ABSTRACT

The end-use model SIMDEUM for residential water demand has been extended to also incorporate non-residential water demand. The model was developed to predict water demand patterns with a small time scale (1 second) and small spatial scale (at the water meter connection). The end-use model is based on statistical information of users and end uses: data on occupancy; the frequency of use; duration and flow per water-use event; occurrence over the day for different end uses such as flushing the toilet, doing the laundry, washing hands, etc. The model follows a modular approach in that each type of building is composed of functional rooms, such as lodging, restaurant and conference rooms. A functional room is characterised by its typical users and water using appliances. With this approach, for non-residential buildings water demand patterns over the day can be simulated.

The simulation results for an office building, hotel and nursing home were compared to measured water demand patterns on attributes such as peak flow and daily total water use, as well as on the shape of the pattern. The simulation results show a good correspondence to measured water demands.

The end-use model is based on independent statistical information and not on flow measurements. The input parameters are available before any information on annual or daily water use is available; the parameters are not fitted on flow measurements. Therefore, the model is transferable to a diverse range of non-residential water demand types. The model can be applied in the design stage (pre-build), in scenario studies, and in distribution network models.

4.1 INTRODUCTION

In the Netherlands, 72% of total water demand is residential, 12% is for small scale business users and 16% is for large scale business users (Geudens 2008). Small scale business users are defined as non-residential connections that use less than 10,000 m^3/year. These users include farms, shops, offices, campgrounds and holiday resorts, small industries, schools, health care institutions and offices (Baggelaar and Geudens 2008). Large scale business users are defined as connections that use more than 10,000 m^3/year.

---

*Reprinted with adaptations from Blokker, E. J. M., Pieterse-Quirijns, E. J., Vreeburg, J. H. G., and van Dijk, J. C. (In Press). "Simulating non-residential water demand with a stochastic end-use model." *Journal of Water Resources Planning and Management*, doi:10.1061/(ASCE)WR.1943-5452.0000146 with permission from the copyright holders, ASCE.

For residential water demand two types of demand models are available, viz. the Poisson Rectangular Pulse (PRP) model (Buchberger et al. 2003) and the end-use model SIMDEUM (Chapter 3). The end-use model SIMDEUM is a physical model that requires information on water using appliances and users. This model can be used in pre-build design problems and in scenario studies. The PRP model requires a measurement campaign to establish the values of the PRP parameters. The PRP model works well as a descriptive model; however, it does not provide insight into the factors that influence the water demand (Chapter 5).

For non-residential water demands, no demand models are available on small temporal and spatial scales. The water demand of large scale business users is measured by most of the Dutch water companies on relatively small time scales (5 minutes). These water demand patterns are user specific and may be relatively constant. A stochastic water demand model would thus not be favourable. Also, the need for a water demand model is not very large because measurements are usually available. For small scale businesses there is a need; most prevalent are offices, hotels and nursing homes (Pieterse-Quirijns et al. 2009). Because the small scale business users are diverse in type and size, the PRP model would require measurements for all these types and sizes. The end-use model, however, offers a perspective on a relatively easy way to construct a water demand model.

In this paper an end-use model for office buildings, hotels and nursing homes is developed and discussed. The model results are compared to measurement data.

4.2 METHODS AND MATERIALS – MODEL DEVELOPMENT

4.2.1 Basic model

The end-use model produces a water demand pattern which can be described by the following equations (Chapter 3):

$$Q = \sum_{k=1}^{M} \sum_{j=1}^{N} \sum_{i=1}^{F_{jk}} B(I_{ijk}, D_{ijk}, \tau_{ijk}) \tag{4.1}$$

$$B(I_{ijk}, D_{ijk}, \tau_{ijk}) = \begin{cases} I_{ijk} & \tau_{ijk} < T < \tau_{ijk} + D_{ijk} \\ 0 & elsewhere \end{cases} \tag{4.2}$$

Here, i, j and k are indices; k counts all end uses from 1 to M, j counts all users from 1 to N, i counts all water use occurrences per end use from 1 to F_{jk} (the frequency of use for user j and end use k). D is the pulse duration (in seconds), I is the pulse intensity (flow in L/s) and τ is the time at which the tap is opened. Thus, D_{ijk} is the duration for end-use type k for user j and occurrence i. $B(I, D, \tau)$ is a block function, which equals I at time τ to $\tau + D$ and 0 during the rest of the day. The summation is done for all (M) available end uses, all (N) users and per end use for the frequency of use (F); this leads to the total water demand pattern Q (L/s) over the day.

In the case of non-residential water demand Q of Eq. 4.1 is determined per functional room and then summed.

$$Q = \sum_{h=1}^{L} Q_h \tag{4.3}$$

Here h is an index that counts the functional rooms 1 to L.

All parameters are described by probability distribution functions which are different for each end use.

4.2.2 The functional rooms (*h*)

For office buildings, hotels and nursing homes the following functional rooms were identified: meeting area, lodging, restaurant, fitness room and other (e.g. technical room). Each functional room has its own end-use types and users (Table 4.1 and Table 4.2). The meeting area contains toilets and coffee machines and users are employees and visitors. The functional room denoted by 'lodging' is a hotel room or a room in a nursing home. These functional rooms involve a water closet (WC), hot and cold water tap, shower and/or bath and users with water related behaviour which is more or less residential. The difference between lodging in hotels and nursing homes is that in hotels, the water use is determined by the hotel guests; in a nursing home, the water use is determined much more by the care takers than by the residents themselves. The restaurant can be a cafeteria in an office building or nursing home or a full service restaurant in a hotel and it is defined by a kitchen tap and dishwashers. In the kitchen, the users are the kitchen personnel. The fitness room in a hotel or office building contains a shower and potentially the same users as the meeting room, but the frequency of use will be lower than for residential showers. The category 'other' includes industrial washing machines, a hot water tap to fill buckets for cleaning and other services. Here, the users are merely 'virtual': their water related behaviour is not connected with individual choices, but to the purpose of water use.

Table 4.1 Users in functional rooms

Functional room	**Type of users (and reason for variable occupancy)**		
	Office	**Hotel**	**nursing home**
Meeting area	Employees (fulltime/part-time working), visitors	Guests (meeting, conference, theatre), employees	Visitors, employees
Lodging	-	Hotel guests (tourists, business people)	Residents
Restaurant	Kitchen personnel	Kitchen personnel	Kitchen personnel
Fitness room	Employees using the fitness room	Hotel guests using the fitness room	-
Technical / other	Number of buckets filled for cleaning, washing machines, …		

4.2.3 The end uses (*k*)

Based on the residential water demand model, eight main types of end uses are defined at the fixture or appliance level, viz. WC, shower, washing machine, dishwasher, kitchen tap, bathroom tap, bathtub and outside tap. For a model of non-residential water demand these types are extended with a coffee machine. The number of installed appliances in a functional room is also of importance. This is denoted by the penetration rate of the different types; for example 6 WCs and 6 cold water taps in the functional room 'meeting area'. For non-residential water use, the end-use presence is not statistically determined, but follows from a fixed input parameter.

Table 4.2 Frequency, duration and intensity for several types and subtypes of end uses in different functional rooms, average (μ) and probability distribution function (pdf)

Functional room	End-use type / subtype		Frequency (day^{-1})*	Duration		Intensity (L/s)	
			μ	μ	Pdf	μ	pdf
Meeting area	WC	6-litre cistern	4.0	2.4 min†	N.A.	0.042	N.A.
		9-litre cistern		3.6 min†			
		Urinal‡		9 s		0.167	
	Tap	Washing hands	4.5	16 s	N.A.	0.083	N.A.
	Coffee machine§		8.0	4.8 s	N.A.	0.042	N.A.
Lodging	Bathtub	120 litres	0.2	10 min	N.A.	0.200	N.A.
	WC	6-litre cistern	4.0	2.4 min†	N.A.	0.042	N.A.
		9-litre cistern		3.6 min†			
	Tap	Washing and shaving	6.0	40 s	Log-normal	0.083	Uniform
		Brushing teeth		15 s			
		Other		45 s			
	Shower	Normal	0.2–0.8**	8.5 min	χ^2	0.142	N.A.
		Water saving				0.123	
		Comfort				0.365	
Restaurant	Dish washer	Brand and type	var.††	Specific dishwashing pattern (2–14 L per cycle, ca. 50 s.)			
	Kitchen tap	Cold / short	var.††	15 s	Log-normal	0.167	Uniform
		Cold / long		60 s		0.250	
		Hot / short		15 s		0.167	
		Hot / long		60 s		0.25	
Fitness room	Shower	Normal	0.8	8.5 min	χ^2	0.142	N.A.
		Water saving				0.123	
Technical / other	Cleaning	Filling bucket	var.‡‡	30 s	N.A.	0.500	N.A.
	Washing machine	Brand and type	var.	Specific washing pattern (9–200 L per cycle)			

**Frequency is Poisson distributed.*
†With a water saving option the duration is reduced to 50% of the original value for 80% of flushes.
‡Frequency of use for men only, division toilet/urinal is 1/3.
§Also tea, water, etc. Water use is the filling of the reservoir.
***0.2 at nursing homes; 0.8 in hotels.*
††Variable: frequency for dishwasher (per guest) depends on cutlery and crockery used in restaurant and on size of dishwasher; frequency for kitchen tap depends on number of dishes that are prepared per guest.
‡‡Frequency for cleaning is number of buckets being filled; this relates to floor space.

For each type of end use there are different subtypes which constitute the end use. For a WC, this can be an old-fashioned toilet with a large cistern of 9 to 12 litres or a new one of only 6 litres with a water-saving possibility of flushing only 3 litres. For the tap, a subtype is defined by its use, e.g., washing hands and brushing teeth. For a model of non-residential water demand other subtypes than for the residential water demand model come into view, such as industrial dishwashers. The penetration rate of the subtypes in a functional room is the fraction of appliances that constitute the type (e.g., type of toilet) or the use per type (e.g., washing hands for the tap in functional room *lodging*).

4.2.4 Users (*j*)

Users are key in an end-use model because they induce the end uses. Users are defined per functional room. Unlike for the residential end-use model, users are not defined as individuals, but as a group. In the model, properties of a non-residential user group are its diurnal pattern (see paragraph 4.2.7) and the number of male and female individuals. A gender distinction was made because it influences toilet use, specifically when a functional room has urinals and toilets. The diurnal pattern work s as the probability distribution function of water use events over the day for the user group and is related to the presence of users over the day. Table 4.1 summarises for office buildings, hotels and nursing homes, the functional rooms and the types of users.

The number of people that are present in the office on any one day will vary. In the Netherlands, 92% of men work full time (40 hours per week), the remaining 8% works on average 17.4 hours (standard deviation 8.1 h). Women on average work 25.5 hours (standard deviation 10.9 h; Bosch et al.). The working hours per week follow a normal probability distribution (Pieterse-Quirijns et al. 2009). In the simulation, the total occupancy per day is determined as follows:

- Of the male employees 92% is present as a base value. For the remaining 8%, first the number of working hours per week is determined ($\mu = 17.4$ $\sigma = 8.1$). Next, for the subjects of whom the part time factor (the number of working hours per week / 40) exceeds a random number r ($\mu = 0.6$; $\sigma = 0.1$) the presence for the simulated day is set to 1, for the others it is set to 0. Total presence is the sum of the base value and the number of part timers present.
- For the female employees the number of working hours per week is determined ($\mu = 25.5$; $\sigma = 10.9$). Next, for the subjects of whom the part time factor exceeds r the presence for the simulated day is set to 1, for the others it is set to 0. Total presence is the sum of the number of part timers present.
- The number of visitors is determined from a normal probability distribution (μ = average number of visitors; $\sigma = 0.2*\mu$) and divided by two, because most visitors do not stay all day. The ratio between male and female visitors is the same as the ratio for employees.

The number of hotel guests that are occupying a hotel room, during a simulation, is determined in the following way:

- A room is occupied or not – the mean occupancy rate is determined from a beta probability distribution. The beta probability distribution is used because it is by definition between 0 and 1; the parameters for the beta probability density function can be determined from the mean and standard deviation. The mean occupancy is an input parameter and can be different for weekdays and the weekend; the standard deviation is assumed to be equal to one quarter of the mean value.
- An occupied room is used by either one person or two persons. In business hotels during the week ca. 60% of the occupied rooms is used by only one person, while in the weekend and in tourist hotels ca. 60% of the occupied rooms accommodate two people (Pieterse-Quirijns et al. 2009).
- the number of visitors is assumed to be a fixed percentage of the lodging guests.

The number of residents in a nursing home is assumed to be more or less stable. Given the fact that there are waiting lists for nursing homes, the occupancy is assumed to be 100%. For other types of users, such as kitchen personnel, there is even less information available on the variability of their presence. Therefore, in the water demand model the number of users in the restaurant, fitness room and technical room is assumed to be fixed number.

4.2.5 The frequency of use (*f*)

For each end use, the frequency of use was determined; either based on the information of the frequency for residential water demand or on an educated guess. The frequency is the number of uses per person per day. Note that in the model the kitchen tap, dishwasher and washing machine, are operated by virtual users and the frequency is determined by other factors. The frequency of use for the dishwasher in a restaurant, e.g., depends on the number of guests and the amount of cutlery and crockery that needs to be washed which is different for each type of meal that is being served.

The frequency of use is described by a discrete statistical distribution, i.e. the Poisson distribution. For non-residential water demand, there is no specific statistical information available for the frequency of use. The Poisson distribution is assumed for all end uses because it was applicable for the frequency of most residential water demand end uses. Table 4.2 summarises for all end uses how f is described.

4.2.6 The pulse intensity (*I*) and pulse duration (*D*)

For each end use, the flow (in L/s) and duration (in s) is based on information of residential end uses, and of technical information on water-using appliances as collected from installation guides and an internet search (Pieterse-Quirijns et al. 2009). The duration of taking a shower is typically determined by the user, while the duration of filling the toilet cistern is determined by the volume of the cistern and the inlet flow which is related to the indoor plumbing. The duration of taking water from a bathroom or kitchen tap, which is determined by the user, can be described using a lognormal distribution function. The same statistical distributions as for the residential water demand end uses are assumed in this model. Table 4.2 summarises for all end uses how I and D are described.

4.2.7 The diurnal pattern, time of water use (τ)

There is no survey available that addresses for all end uses the time of the day when a water-use event is likely to take place. For the residential end-users it was shown that the time of day of water use is strongly related to whether people are at home or not and if they are asleep, getting up or preparing for bed (Chapter 3). This relation is also used for the functional room lodging in a hotel. For hotel guests, a sleep-wake-rhythm is applied (Table 4.3) that differs from residential users' diurnal pattern. Different diurnal patterns are applied for tourists and business travellers. For the non-residential users such as employees or restaurant guests and the nursing home residents the diurnal patterns are related to presence during e.g. lunch hours, defined by t1–t4 (Table 4.4). Time t1 and t4 denote the edges of people's presence; between t2 and t3 a higher probability of use occurs. These parameters were derived from specific surveys (Pieterse-Quirijns et al. 2009), see also Section 4.6.

A diurnal pattern for a residential user is constructed with random picks from the normal distribution. This pattern is converted to a probability density function (PDF) of water-use events in the following way (Chapter 3):

- During the sleeping hours, the total volume of water use is estimated at 1.5% of the total daily demand.
- During absence, the probability of (the start of) water use is set to zero.

- During the half hour after getting up and returning home and the half hour before leaving home and going to bed, peak hours are assumed and the fraction of the total demand occurring in this period is estimated at 65%.
- The rest of the water use then corresponds to 33.5% of the daily water use.

Table 4.3 Statistics for residential diurnal patterns (Pieterse-Quirijns et al. 2009)

		Tourist	Business traveller	Elderly in retirement home
Time of getting up	μ	8:00	7:00	8:00
	σ	1:30	1:00	1:00
Time of leaving the house	μ	10:00	8:30	13:00
	σ	2:00	1:00	3:00
Duration of being away	μ	14.0 h	9.0 h	10.0 h
	σ	1.0 h	2.0 h	4.5 h
Duration of sleep	μ	8.0 h	8.0 h	8.0 h
	σ	1.0 h	0.5 h	1.0 h

The cumulative distribution function (CDF) is the cumulative sum of the PDF and is scaled to be between zero and one. Because the times of waking up, leaving the home, returning home and going to bed vary (Table 4.3) the probability of water-use events varies as well. A random time of water use (τ) can be retrieved from the CDF. An example is shown in Figure 4.1a where a random number equal to 0.45 means that τ for the water-use event is 18:15.

For the group of non-residential users the diurnal pattern is constructed in a similar manner from given μ and σ of t1 to t4:

- Around t1 and t2, following a normal distribution, a PDF is constructed which equals 0 at $\mu - 3\sigma$ and is at its maximum at $\mu + 3\sigma$.
- Around t3 and t4, following a normal distribution, a PDF is constructed which is at its maximum at $\mu - 3\sigma$ and equals 0 at $\mu + 3\sigma$.

The cumulative distribution function (CDF) is the cumulative sum of the four PDF and is scaled to be between zero and one. A random time of water use (τ) can be retrieved from the CDF. An example is shown in Figure 4.1b where a random number equal to 0.3 means that τ for the water-use event is 11:15.

4.3 METHODS AND MATERIALS – THE SIMULATION

Once the statistical information on users and water outlets per functional room are put into the model, the simulation can be run. The simulation scheme is depicted in Figure 4.2.

First, for each simulated functional room, the gender and occupancy (§4.2.4) and diurnal patterns of the occupants (Table 4.3 and Table 4.4) are determined. Next, for each simulated functional room it is determined what end uses ($k = 1$ to M) and subtypes are present and how many. Then, per end use k (in random order), for all users ($j = 1$ to N) in the functional room, the frequency of use (F_{jk}) is determined from the Poisson distribution function (Table 4.2). The frequency of use is divided over the total of number of appliances in a group (e.g. all ladies WC's). After that, for all occurrences ($i = 1$ to F_{jk}), the

Table 4.4 Statistics for non-residential diurnal patterns (Pieterse-Quirijns et al. 2009)

		Office employee	Office cafeteria personnel	Office cleaner	Hotel conference guest	Hotel restaurant personnel	Hotel cleaner	Nursing home residents	Nursing home employees + visitors
t1	μ	8:00	8:00	17:45	8:00	7:30	9:00	6:30	6:00
	σ	1:00	1:00	0:05	1:00	1:00	1:00	1:00	2:00
t2	μ	12:15	12:30	-	11:00	10:00	-	7:30	8:00
	σ	0:30	0:30	-	1:00	1:00	-	0:30	0:30
t3	μ	13:00	13:30	-	20:00	20:00	-	22:00	21:00
	σ	1:00	1:00	-	1:00	1:00	-	1:00	0:30
t4	μ	18:00	15:00	18:15	23:00	22:30	13:00	22:30	22:00
	σ	1:00	0:15	0:05	1:00	0:30	1:00	1:00	2:00

duration and intensity are determined from the appropriate PDF (Table 4.2) and when during the day (τ) the water-use event occurs (from the diurnal PDF, Figure 4.1). When for end use k, user j and occurrence i the time τ_{ijk} is established, the time $\tau_{ijk} - D_{ijk} < \mathrm{T} < \tau_{ijk} + D_{ijk}$ is blocked for end use k, user $j+1$ to N and occurrence $i+1$ to F_{jk}. This ensures that there is no overlapping use by multiple users of one end use, e.g., when only a single shower is present, the use of it by one person cannot coincide with its use by a second person. However, user j could potentially draw water from multiple end uses (k and $k+1$) at the same time. This is not actively blocked, but seems to occur seldom in residential water use and is of less importance with the (virtual) non-residential users. The sum of all water demands is the water demand pattern of the functional room; the sum of the water demands of all functional rooms is the water demand pattern of the non-residential building.

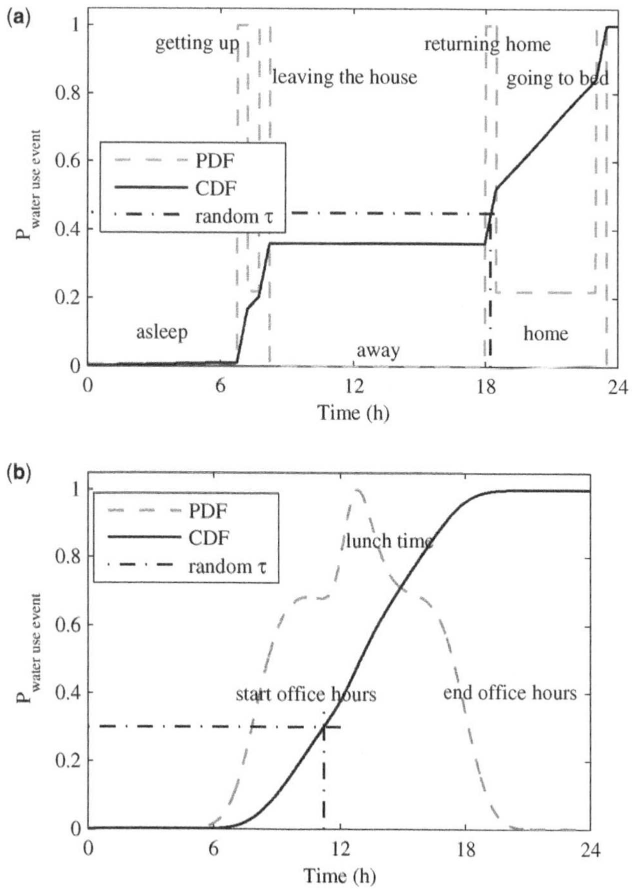

Figure 4.1 An example of drawing a random time (τ) from the CDF of water use for a) a business traveller with a residential diurnal pattern and b) an office employee with a non-residential diurnal pattern

The output of one simulation run is one water demand pattern. By doing more simulations, multiple patterns (days) of a specific building are found. Due to the stochastic nature of water demand patterns, this single simulated water demand pattern is only one possible outcome. With a Monte Carlo simulation, many simulation results are obtained and a statistical analysis can be made on the results. The end-use model was programmed in MATLAB® to run the simulations.

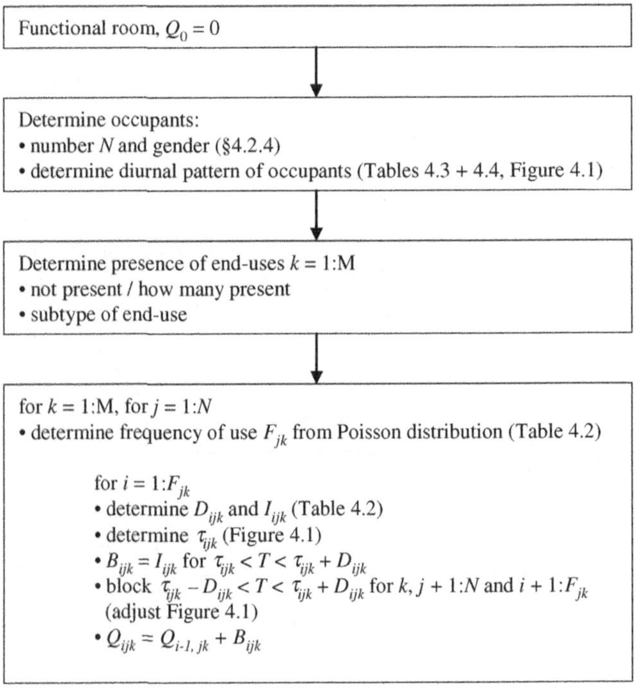

Figure 4.2 Simulation scheme

4.4 METHODS AND MATERIALS – MODEL VALIDATION

To validate the water demand model, the results are compared to water demand measurements on several characteristic parameters. Since the output of the model is stochastic in nature, the frequency distribution of the characteristics is used in the comparison. To quantify the validation, some statistic measures were used.

For the residential water demand model eight different characteristic parameters were used which are suitable for different time scales and spatial scales (Chapter 3). For the validation of the non-residential water demand model of an office building, hotel and nursing home, the characteristics Q_{max} (the maximum flow for different time scales), V (the total volume per day in m^3), Q_{day} (the demand pattern over the day in m^3/h) and C_d (the normalised multiplier pattern Q_{day}) are used.

The parameters Q_{max} and V are presented as cumulative frequency distributions. The goodness-of-fit for the cumulative frequency distributions is determined with ME, RMSE and R^2 evaluated at each 10th integer percentile between 10 and 100 (10, 20, 30, … 100). Q_{day} is a pattern; the goodness-of-fit is expressed as ME, RMSE and R^2 of Q_{day}. The ME and RMSE are presented as absolute values and as percentages of the means of the measurements.

4.5 METHODS AND MATERIALS – SENSITIVITY ANALYSIS

A simple sensitivity analysis was done to determine which functional room has the largest influence on the total water use. The information of the most contributing functional room has to be more reliable than the information of functional rooms that contribute only marginally to the total water use. Pieterse-Quirijns et al. (2009) did a more extensive sensitivity analysis for three more office buildings, three more hotels and two more nursing homes (without assisted living).

4.6 RESULTS – MODEL VALIDATION

To establish the performance of the water demand model, the results are compared to water demand measurements. Demand measurements were collected for office buildings, hotels and nursing homes from several Dutch water companies. Specific input parameters for these buildings were gathered from surveys. These surveys enquired after the number of employees and visitors, the number of toilets etc. In total, water demand measurements and surveys of four office buildings and four hotels were collected; for the nursing homes only one survey was returned and measurements of three other nursing homes were supplied by a water company. As an illustration, the results of one office building, one hotel and one nursing home are presented here.

4.6.1 Office building

The particular office building that was simulated and validated is an office of a Dutch water company. Measurements were available on a 1 minute time basis for two weeks (10 weekdays) in November 2008.

In this office 284 men and 189 women are employed; on average 50 visitors per day are received. The office contains 38 toilets (7 urinals, 18 toilets for men, 13 toilets for women), 4 showers and 15 coffee machines. The kitchen is open for lunch: 125 people make use of that. The kitchen contains one dishwasher (frequency is set to 0.3 per visitor for simple meals) and some kitchen taps for hot and cold water. The office is cleaned every day at the end of the day (after 17:00 hours) by 12 people filling 12 buckets of warm water and 12 buckets of cold water at the start of their shift.

For the simulation 50 weekday patterns were constructed for this office building. The maximum flow (Q_{max} per 5 minutes and per hour), the volume (V per day) and the pattern over the day (Q_{day}) for measured and SIMDEUM water demand patterns are compared in Figure 4.3, and Table 4.5.

4.6.2 Hotel

The particular hotel that was simulated is a hotel that also provides conference rooms and a theatre. Measurements were available on a 5 minute time basis for 30 days in June 2006.

The hotel has 112 rooms with an average occupancy of 80% during the week (60% one person, 40% two persons) and 40% during the weekend (40% one person, 60% two persons). The room contains a toilet, a tap for cold and hot water, a shower and a bath.

For the conference rooms and theatre, the hotel has 120 toilets (40 urinals, 40 toilets for men, 40 toilets for women), 69 washstands and 7 coffee machines. Conference visitors ($\mu = 350$, $\sigma = 100$) are present during the day; theatre visitors ($\mu = 350$, $\sigma = 100$) are present during the evening.

The hotel serves breakfast, lunch and dinner for on average 92, 150 and 100 guests respectively. In the kitchen, water is used for food preparation and dishwashing. The kitchen contains one dishwasher; its frequency of use per guest is estimated at 0.2 for breakfast, 0.3 for lunch and 0.5 for dinner.

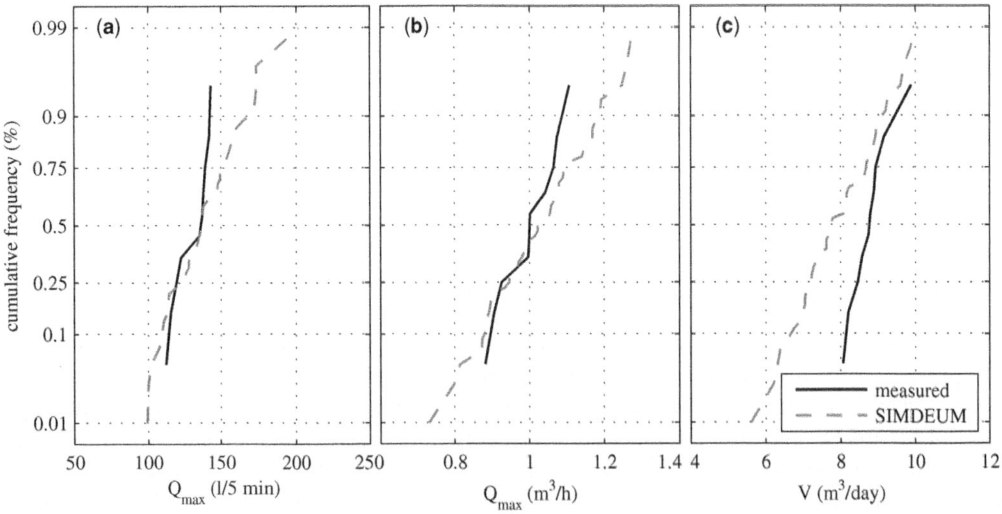

Figure 4.3 Office building: comparing measurements (10 weekdays at 1 min time scale) and simulations (50 days at 1 s time scale); a) Q_{max} per 5 min, b) Q_{max} per h, c) V per day

Table 4.5 Statistics of comparison between measured and simulated water demand from Figure 4.3 and Figure 4.4. All values are statistically significant

		ME		RMSE		R^2
Q_{max}	(l/5 min)	11.5	(8.7%)	22.4	(17.0%)	N.A.
Q_{max}	(m^3/h)	0.0	(4.0%)	0.1	(7.1%)	N.A.
V	(m^3/day)	−0.8	(−8.5%)	1.0	(10.9%)	N.A.
Q_{day}	(m^3/h)	−0.0	(−10.1%)	0.1	(25.0%)	0.93

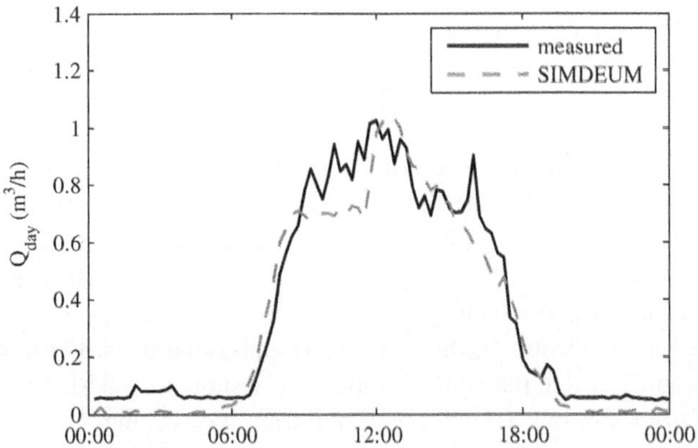

Figure 4.4 Office building: comparing average Q_{day}, at a time scale of 15 min, of measurements (10 weekdays) and simulations (50 days)

The hotel cleaning is done in the morning by ca. 20 cleaners, taking hot and cold water for cleaning the hotel rooms and conference rooms.

For the simulation 50 patterns were constructed for this hotel. The maximum flow (Q_{max} per 5 minutes and per hour), the volume (V per day) and the pattern over the day (Q_{day}) for measured and SIMDEUM water demand patterns are compared in Figure 4.5, Figure 4.6 and Table 4.6.

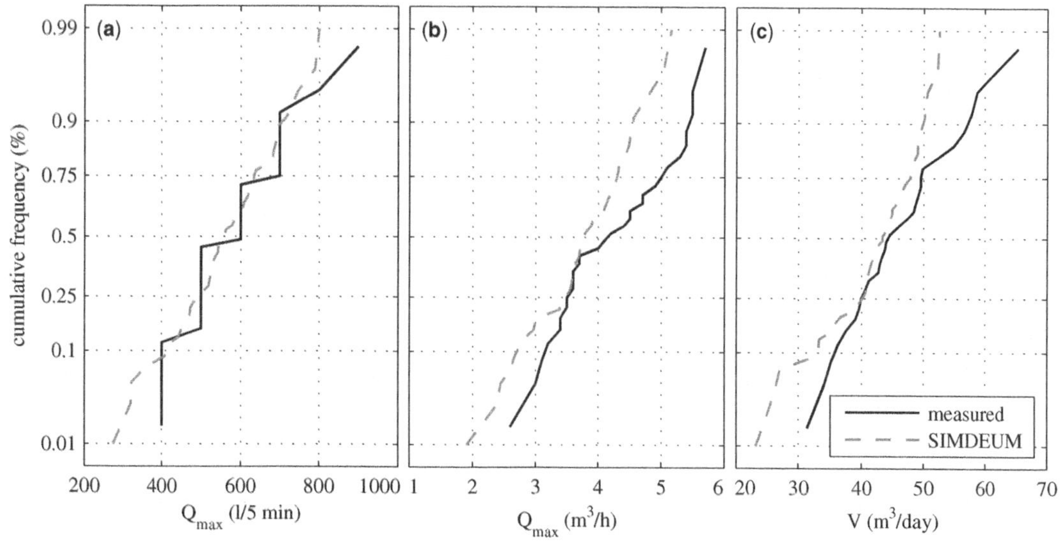

Figure 4.5 Hotel: comparing measurements (30 days at 5 min time scale) and simulations (50 days at 1 s time scale); a) Q_{max} per 5 min, b) Q_{max} per h, c) V per day

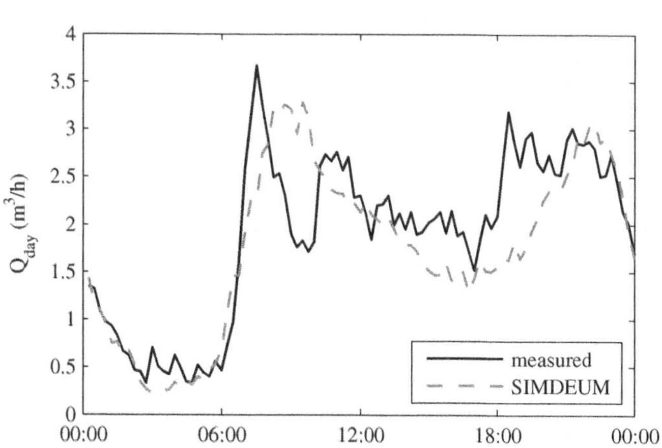

Figure 4.6 Hotel: comparing average Q_{day}, at a time scale of 15 min, of measurements (30 days) and simulations (50 days)

Table 4.6 Statistics of comparison between measured and simulated water demand from Figure 4.5 and Figure 4.6. All values are statistically significant

		ME		RMSE		R^2
Q_{max}	(l/5 min)	−10.8	(−1.8%)	41.2	(6.9%)	0.91
Q_{max}	(m^3/h)	−0.4	(−10.0%)	0.6	(12.6%)	0.62
V	(m^3/day)	−3.6	(7.6%)	5.4	(11.3%)	0.63
Q_{day}	(m^3/h)	−0.1	(−7.4%)	0.5	(28.6%)	0.61

4.6.3 Nursing home

The particular nursing home that was simulated has 110 single rooms in the nursing home and 80 apartments for assisted living. Measurements were available on a 5 minute time basis for 30 days in June 2006. No survey was filled out; specific information was estimated from the nursing home's website and a survey of another nursing home.

The apartments for assisted living house 1 or 2 people over 65 years of age. They have a residential water demand through toilet, shower, kitchen and bathroom tap and washing machine.

In the nursing home a maximum of 350 people are present during the day. This is a combination of personnel and visitors (240) and residents (110). For personnel and visitors 40 toilets and 40 washstands are available and 10 coffee machines. The single rooms each have a toilet, a washstand and a shower. Each floor (5 in total) has a small kitchen to prepare meals and do some washing up which uses a small amount of water. The restaurant has a large dishwasher which runs ca. 120 times per day. Extra water use is induced for cleaning (ca. 20 cleaners during the entire working day), hair dressing facility (ca. 40 water use events per day), washing machines (ca. 6 runs per day) and bedpan cleaning.

For the simulation 50 patterns were constructed for this nursing home plus assisted living. The maximum flow (Q_{max} per 5 minutes and per hour), the volume (V per day) and the pattern over the day (Q_{day}) for measured and SIMDEUM water demand patterns are compared in Figure 4.7, Figure 4.8 and Table 4.7.

4.7 RESULTS – SENSITIVITY ANALYSIS

Figure 4.9 shows for an office, a hotel and a nursing home how much each functional room contributes to the total water use.

4.8 DISCUSSION

The simulations with the presented non-residential SIMDEUM and the comparison with measured non-residential water demand show the following. For office buildings, the maximum flow and the flow pattern over the day typically is predicted well by SIMDEUM (Pieterse-Quirijns et al. 2009), i.e. ME and RMSE $< 30\%$, and $R^2 > 0.7$. For hotels, also, the maximum flow and the flow pattern over the day typically is predicted well by SIMDEUM. This is the case for this particular hotel with conference rooms and theatre (which has a relatively high water use in the evening) as well as for more common tourist hotels (Pieterse-Quirijns et al. 2009). For nursing homes, the maximum flow per 5 minutes and the flow pattern over the day typically is predicted well by SIMDEUM. This is particularly remarkable, since only very limited information of the nursing homes was available, because no surveys were filled in. The diurnal pattern for nursing homes without assisted living looks different from Figure 4.9 in that the low

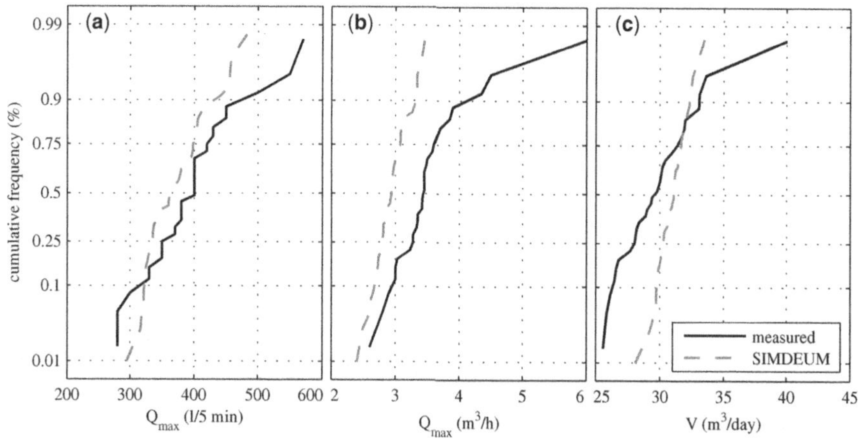

Figure 4.7 Nursing home: comparing measurements (30 days at 5 min time scale) and simulations (50 days at 1 s time scale); a) Q_{max} per 5 min, b) Q_{max} per h, c) V per day

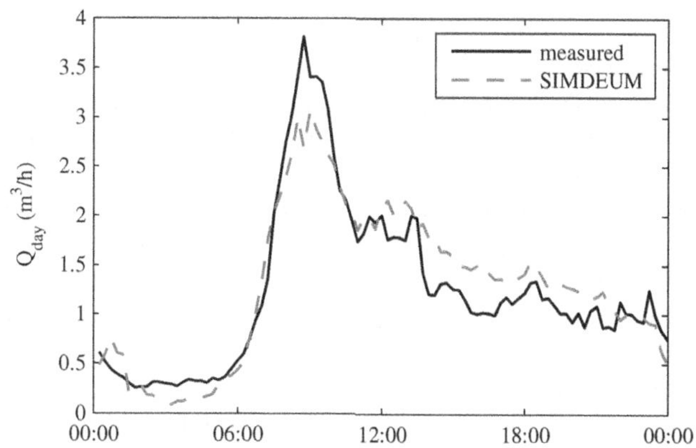

Figure 4.8 Nursing home: comparing average Q_{day}, at a time scale of 15 min, of measurements (30 days) and simulations (50 days)

Table 4.7 Statistics of comparison between measured and simulated water demand from Figure 4.7 and Figure 4.8. All values are statistically significant

		ME		RMSE		R^2
Q_{max}	(l/5 min)	−29.6	(−7.2%)	38.1	(9.3%)	0.71
Q_{max}	(m^3/h)	−0.7	(−19.2%)	1.0	(26.8%)	N.A.
V	(m^3/day)	−0.7	(2.3%)	2.9	(9.3%)	0.47
Q_{day}	(m^3/h)	0.1	(4.3%)	0.3	(22.9%)	0.88

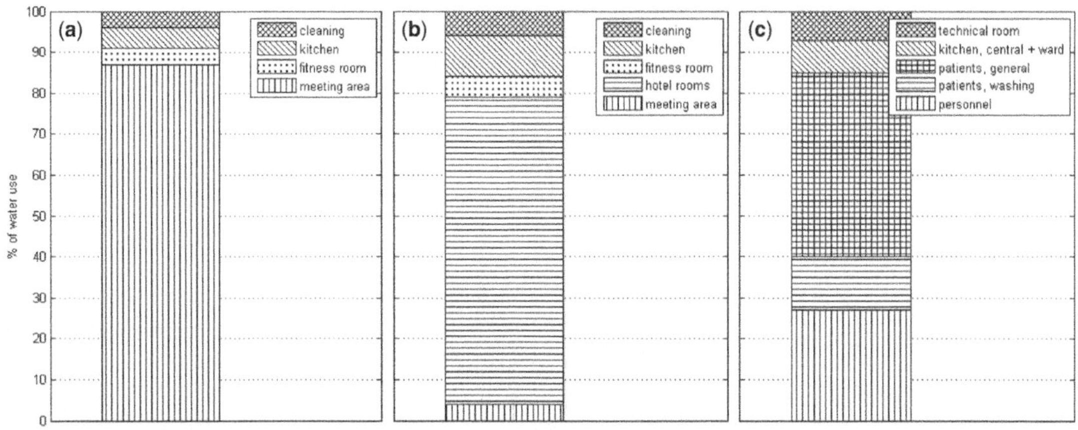

Figure 4.9 Percentage of total water use per functional room for a) an office building; b) a hotel and c) a nursing home

night use starts earlier (before midnight); this is the case for both measured and simulated water demand (Pieterse-Quirijns et al. 2009).

The sensitivity analysis (Figure 4.9) showed that in office buildings, the meeting area is the main contributor; more specifically the toilets in the meeting area. Thus, it is important to have reliable information on the toilet use: how many people are employed, what is the frequency of use, how much water is used per flush, is a water saving option available and how often is this water saving option used? In hotels, the main contributor is the sum of hotel rooms. Thus, it is important to have reliable information on the rooms: how many rooms are present, what is the occupancy, what type of shower and toilet are installed (i.e. is a water saving device is installed or a luxurious shower)? The water use in the presented nursing home is not determined by a single dominant functional room. The water use is largely determined by the toilet use of personnel and residents and by the showering of residents.

The end-use model uses less stochastic information for the non-residential water demand than for the residential demand. For example, the number of water using appliances is a fixed number and the users are simulated as a group of users rather than as individual users. The diurnal pattern for the probability of a water use events is specific per functional room, and is valid for all users alike. This approach of using a specific diurnal pattern appears to work well. The parameters for start and end time of user presence and higher probability of water use were collected from only a few surveys and are not yet generalised for all functional rooms. Nevertheless, the end-use model works well. The assumptions on the frequency of use, duration and flow of water use events and the probability of when such an event is most likely to occur thus seem to be estimated well. The theoretical basis of the end-use model ensures good simulation results. This is very promising for other types of non-residential buildings, e.g. a penitentiary, that share some functional rooms with the building types described in this paper.

Compared to the end-use model for residential water demand, the end-use model for non-residential water demand was not fully developed. The information on non-residential water use is less generic because little information could be found in literature and only a few project specific surveys were available. Also, the validation of the non-residential water demand model was only done for a limited number of buildings, which also filled out the surveys. More validation is recommended for non-residential buildings for which no filled out surveys were available.

4.9 CONCLUSIONS

The end-use model SIMDEUM, that was developed to predict residential water demand, can also be applied to predict non-residential water demands. The explanation for this flexibility is that information on the water using appliances and water users is input to the model. The model uses functional rooms that can be used in different types of non-residential buildings, not just the presented office, hotel and nursing home.

The water use in office buildings is mainly determined by flushing of toilets. Thus, it is important to have reliable information on the toilet use. The water use in hotels (both total use and diurnal pattern) is mainly determined by the use in the hotel rooms. Thus, it is important to have reliable information on the number and occupancy of the rooms and the water using appliances in the room. The water use in the presented nursing home is not determined by a single dominant functional room.

First results show that the simulation results are in good agreement with measured water demand patterns. This is true for a range of time scales (from 5 minutes to one hour) and a spatial scale of a single office, hotel and nursing home. The assumptions on the frequency of use, duration and flow of water use events and the probability of when such an event is most likely to occur thus seem to be estimated well. The results are promising; the model may also be applied for other types of non-residential buildings.

Chapter 5

Comparison of two approaches of residential water demand modelling*

ABSTRACT

There is growing interest in modelling water demands on short time scales (as brief as one second) and small spatial scales (typically single homes). Buchberger et al. (1996, 2003) have developed the Poisson Rectangular Pulse (PRP) model for this purpose, based on flow measurements. Blokker et al. (Chapter 3) have developed an end-use model, SIMDEUM, which is based on statistical information from end uses and does not require any flow measurements. SIMDEUM was developed and validated for Dutch water use. In this chapter, the approaches of the PRP model and SIMDEUM are compared with each other and with measured indoor water demands from 21 homes in Milford, Ohio. Both models compare well to the measurements; the PRP model works better in simulating the cumulative flows of a sum of 20 homes, SIMDEUM works better in simulating the flows of a single home.

5.1 INTRODUCTION

There is growing interest in modelling water demands on short time scales (as brief as one second) and small spatial scales (typically single homes). One of the main drivers for this is the growing interest in water quality modelling (Chapter 2). Buchberger et al. (1996; 2003) have developed the Poisson Rectangular Pulse (PRP) model for this purpose. Blokker et al. (Chapter 3) have developed the end-use model SIMDEUM. The objective of this chapter is to compare the approaches and the results of both water demand models.

SIMDEUM so far has only been demonstrated with Dutch data and compared with Dutch flow measurements. For this study, SIMDEUM was applied to US data and was compared with the measured Milford indoor water demands of 1997 (Buchberger et al. 2003). For this purpose, end-use parameters from Milford (in 1997) were collected from the literature to be used as an input into SIMDEUM. By comparing SIMDEUM to the Milford measurements, SIMDEUM is for the first time validated for a region outside the Netherlands. This is a second objective of this chapter.

The PRP model was not applied to the Dutch situation. And thus, the PRP model and SIMDEUM were not mutually compared to Dutch flow measurements. PRP parameters were obtained from measurements in

*Parts of this chapter were based on: Blokker, E. J. M., Buchberger, S. G., Vreeburg, J. H. G., and van Dijk, J. C. (2008). "Comparison of water demand models: PRP and SIMDEUM applied to Milford, Ohio, data." WDSA 2008, J. E. van Zyl, A. A. Ilemobade, and H. E. Jacobs, eds., Kruger National Park, South Africa, 182–195.

the USA (Buchberger et al. 2003), Italy (Guercio et al. 2001), Spain (García et al. 2004) and Mexico (Alcocer-Yamanaka et al. 2006). These measurements all led to different values of the PRP parameters. This means that the PRP parameters for the Netherlands cannot be known without analysing flow measurements. Existing flow measurement data were not suitable to extract the PRP parameters. An extensive measurement campaign to determine the PRP parameters for the Netherlands was not carried out.

5.2 MATERIALS AND METHODS

5.2.1 The Poisson Rectangular Pulse (PRP) model

The Poisson Rectangular Pulse (PRP) model for residential water use assumes that water demands at a single family home behave as a time dependent Poisson arrival process (Buchberger and Wu 1995). When a water use occurs, it is represented as a single rectangular pulse of random duration and random but steady intensity. Buchberger and Wells (1996) found that over 80% of indoor residential water demands occur as single pulses and more complex water use patterns are readily converted to single equivalent rectangular pulses (SERPs). In a subsequent field study, Buchberger and co-workers (2003) measured nearly 365,000 residential water use pulses at 21 homes in Milford, Ohio, during a 214-day period of continuous monitoring (more info in §5.2.3). The timing and statistical properties of these SERPs provided strong support for the PRP hypothesis for residential water use (Buchberger and Lee 1999; Clark et al. 2004).

The PRP model produces a demand pattern which can be described by the following equations:

$$Q = \sum B(I, D, \tau) \tag{5.1}$$

$$B(I, D, \tau) = \begin{cases} I & \tau < T < \tau + D \\ 0 & elsewhere \end{cases} \tag{5.2}$$

with I the pulse intensity (L/s), D the pulse duration (s) and τ the starting time of the pulse. The PRP model for household water use determines the intensity, duration and starting time of each pulse with the help of five parameters, each with a clear physical interpretation: [i] mean and variance of pulse intensity, α and β^2, respectively; [ii] mean and variance of pulse duration, τ and ω^2, respectively; and [iii] hourly arrival rate, $\lambda(k)$, or hourly demand multiplier, $\pi(k)$. In the Milford study in 21 homes over 31 days, a total of 52,072 SERPs were counted; the mean and variance of the pulse duration was determined and the mean and variance of the intensity during busy hours was also determined. Typical values for water pulse intensity and duration, as estimated from the Milford field studies, are summarised in Table 5.1 and Table 5.2. The average demand per home was 563 L/day.

Table 5.1 Typical parameters for PRP model in Milford, Ohio

Term	Symbol	Units	Value, indoor
Mean pulse intensity	α	L/min	8.52
Variance of pulse intensity	β^2	L/min^2	22.21
Mean pulse duration	τ	min	0.75
Variance of pulse duration	ω^2	$(min)^2$	2.25
Mean arrivals per home	λ	1/min	0.0593

Table 5.2 Typical residential demand pattern in Milford, Ohio

No.	Starting time	Ending time	Pattern multiplier π	No.	Starting time	Ending time	Pattern multiplier π
1	0:00	1:00	0.39	13	12:00	13:00	1.02
2	1:00	2:00	0.22	14	13:00	14:00	0.94
3	2:00	3:00	0.12	15	14:00	15:00	0.99
4	3:00	4:00	0.18	16	15:00	16:00	1.15
5	4:00	5:00	0.50	17	16:00	17:00	1.32
6	5:00	6:00	0.99	18	17:00	18:00	1.42
7	6:00	7:00	1.28	19	18:00	19:00	1.47
8	7:00	8:00	1.32	20	19:00	20:00	1.41
9	8:00	9:00	1.36	21	20:00	21:00	1.29
10	9:00	10:00	1.35	22	21:00	22:00	1.14
11	10:00	11:00	1.29	23	22:00	23:00	0.96
12	11:00	12:00	1.19	24	23:00	24:00	0.70

The range of PRP parameter estimates in Table 5.1 are nominal values averaged over a typical day. At any given home, values of the PRP parameters may change from hour to hour. In most cases, hourly variations in PRP parameter values are relatively minor and can be ignored, except for the arrival rate where a strong diurnal pattern is common. Hourly variations in the arrival rate are the primary source of time dependence in the PRP model and the main reason for spatial correlation among demand nodes (Li et al. 2007). The arrival rate is additive across homes. This is a very important and useful property because it implies that the PRP concept for residential water use is scalable.

An interactive C++ computer code called PRPsym was developed to simulate residential water demands that follow a Poisson rectangular pulse process (Li and Buchberger 2003). The PRPsym code generates instantaneous (i.e. second-by-second basis) water demands for each node in a pipe network. The general algorithm to simulate residential water demands with the PRP model is a three-step process as discussed below.

Step 1 - Generate arrival rate and start time of water pulses: The average arrival rate $\lambda(k)$ for the Poisson process during time step k at a demand node is given by

$$\overline{\lambda(k)} = \pi(k) \times \lambda \tag{5.3}$$

where $\pi(k)$ is a user-specified multiplier for time step k (Table 5.2), and λ is the number of average arrivals per home per min (Table 5.2), and is equal to

$$\lambda = \frac{Q}{\alpha\tau} \tag{5.4}$$

with Q equal to 563 L/day = 0.39 L/min and α and τ from Table 5.1.

On average, the multiplier $\pi(k)$ is unity. During high use periods, $\pi(k) > 1$; conversely, during low use periods $\pi(k) < 1$. Since pulse arrivals follow a Poisson process, the inter-arrival times (waiting times between pulses) are exponentially distributed random variables. Therefore, the waiting time between pulse m-1 and pulse m during the kth time step is given by

$$t_m(k) = \frac{-1}{\lambda(k)} \ln[U_m] \tag{5.5}$$

Here U_m is a uniform random variable on the interval [0, 1]. The start time for the mth pulse during the kth time step is

$$T_m(k) = \sum_{i=1}^{m} t_i(k) \quad 0 \leq T_m(k) \leq 1\text{hr} \tag{5.6}$$

Step 2 - Generate random pulse intensities and pulse durations from a lognormal distribution: If X represents a log-normally distributed intensity or duration with known mean μ and variance σ^2, then $Y = \ln[X]$ is a normally distributed random variable with mean α and variance β^2, given by

$$\alpha = \ln(\mu) - \frac{1}{2}\ln\left(1 + \frac{\sigma^2}{\mu^2}\right) \quad \text{and} \quad \beta^2 = \ln\left(1 + \frac{\sigma^2}{\mu^2}\right) \tag{5.7}$$

The normal random variable Y with parameters α and β^2 is simulated by

$$Y = \alpha + z\beta \tag{5.8}$$

where $z \sim N(0,1)$ is a variate from standard normal distribution. Finally, the log-normally distributed pulse intensity or pulse duration X is given by

$$X = e^Y \tag{5.9}$$

Step 3 - Sum (aggregate) the instantaneous concurrent pulses: The simulated pulses with random start times, intensities and durations found in Steps 1 and 2 are aggregated to obtain a sequence of total instantaneous nodal demands. As shown in Eq. 5.1, the total demand at any time is simply the sum of intensities of all concurrent active PRP demands.

For the purpose of this study, the different programming steps were put into MATLAB®. This simplified the possibility of comparing the results of the PRP model with the results from the end-use model.

5.2.2 The End-Use Model SIMDEUM

SIMDEUM is based on the same principle of rectangular pulses as the PRP model; SIMDEUM produces a demand pattern which can be described with equations similar to Eqs. 5.1 and 5.2, with the exception that I, D and τ are dependent on users and end-use types.

Only indoor water use is being considered. This leads to seven types of fixtures, with a subdivision into a number of end uses. All end uses and frequency, duration and intensity per end use for the 21 homes of Milford are summarised in Table 5.3. The data for Table 5.3 were collected from several sources:

- Buchberger et al. (2003) used information from Maddaus (1987). This source provided mainly information on total water use per person per day for different end uses, built up by frequency and product of duration and intensity of appliances. Buchberger et al. (2003) also provide some specific information on the 21 homes at which demand measurements (Sec. 5.2.3) were taken.

Table 5.3 Penetration rate, frequency, duration and intensity – average (μ) and probability distribution function (pdf) – of end uses as input for end-use model of 21 homes in 1997 in Milford, Ohio

Fixture / end-use type		Penetration rate* (%)		Frequency (day^{-1})		Duration		Intensity (L/s)	
		tap	end use	μ	pdf	μ	pdf	μ	pdf
Bath (120 liter)		100	100	0.13	Poisson	10 min	N.A.	0.200	N.A.
Bathroom tap	Washing and shaving	100	33	4.10	Poisson	40 s	Log-normal**	0.08	Uniform
	Brushing teeth		67			15 s			
Dishwasher		75.3	100	0.25	Poisson		Specific dishwasher pattern (4 × 45 sec, 0.17 L/s, 30 L)		
Kitchen tap	Consumption	100	37.5	12.70	Negative binomial (r = 3, p = 0.192)	15 s	Log-normal	0.083	Uniform
	Dishes		25			45 s		0.125	
	Washing hands		25			13 s		0.083	
	other		12.5			48 s		0.083	
Shower		100	100	0.70	Poisson	8.5 min	χ^2	μ = 0.139 σ = 0.062	N.A.
Washing machine		100	100	0.37	Poisson		Specific washing machine pattern (4×3.3 min, 0.19 L/s, 152 L)		
WC (13.2 L cistern, 3.5 gal)		100	100	5.05	Poisson	1.8 min	N.A.	0.125	N.A.

* *penetration rate: percentage of households that possess a specific appliance or specific subtype (e.g. type of toilet) or it is distribution of use per subtype (e.g. for kitchen tap).*

- Mayer et al. (1999) provide some information for the USA on frequency, duration and intensity (with information of average and standard deviation) for running water at faucets and showers.
- Gleick et al. (2003) provide some information for the USA on intensity of faucets and showers, and on total water use for toilet cistern, dishwasher and washing machine over the years. The information for 1997 was used.
- Unknowns were assumed to be similar to the Dutch data (Chapter 3).

Information on the Milford users was collected from 2000 census data (U.S. Census Bureau 2000) and residents' information (Buchberger et al. 2003). This translated into the input parameters of SIMDEUM as summarised in Table 5.4.

Table 5.4 Composition of households in Milford (U.S. Census Bureau 2000)

		One person households	Two person households	Families with children
Number of people per household		1	2	3.75 (on average)
Number of households (%)		4.8	52.4	42.9
Gender division: Male / Female (%)		50 / 50	50 / 50	50 / 50
Age division (%)	Children (0–12 years old)	0	0	33.1
	Teens (13–18 years old)	0	0	11.7
	Adults (19–64 years old)	50	50	55.2
	Subdivision: % of adults with out-of-home job	Male: 100	Both persons: 100	Both parents: 77.7
			Only male: 0	Only father: 15
		Female: 100	Only female: 0	Only mother: 7.3
			Neither person: 0	Neither parent: 0
	Seniors (>65 years old)	50	50	0

The American Time Use Survey 'ATUS' (U.S. Department of Labor; Bureau of Labor Statistics) gives information about the duration of sleep, time of getting up, time of leaving and returning home. The information can be discerned for different age groups, different days of the week, but not for different states, let alone cities. The US data can be described by normal distributions (Figure 5.1), as could the Dutch data (Chapter 3). Compared to the Dutch data, the US data have a much larger standard deviation. For example, sleeping duration on average is 8 hours (both in the Netherlands and in the USA) and the standard deviation is ca. 1 hour in the Netherlands and 1.8 hours in the USA. The sleep–wake rhythm and working hours of the residents in the specific 21 Milford homes is unknown. At first, an estimate was made based on the Dutch data, the ATUS data and common sense (Blokker et al. 2008). Secondly, the ATUS 1995 data were used, as summarised in Table 5.5. This improved the overall water demand pattern over the day. The data from Table 5.5 are used in this chapter.

5.2.3 The flow measurements of Milford, Ohio

Instrument packs to monitor residential water use were installed on the service line at secure indoor locations in each of the 21 homes at the Milford study site. Residential water consumption including indoor (hot and cold) and outdoor water use as well as leaks were recorded around the clock. The instrument package included a Rustrak Ranger II data logger, 4-29 mA impulse sensor, count and dump input pod, COM-504 modem connection, and 2,400 baud Hayes modem. Data loggers were coupled with an input sensor inserted between the base and register head of a 5/8 inch T-10 Neptune water meter. The input sensor monitored revolutions of a magnet fixed to a positive displacement nutating disc in the measuring chamber of the meter. At 1-second intervals, the sensor transmitted a signal to the data logger where the readings were converted to a flow rate and stored. Most recordings were zero. Water demands were uploaded nightly via dedicated modems linked to a computer at the University of Cincinnati. During the 7-month period, April to November, over 365,000 water demands were registered at the 21 residences on the study site. More details on the process used to transform raw input signals to archived pulses are available in Buchberger et al. (2003).

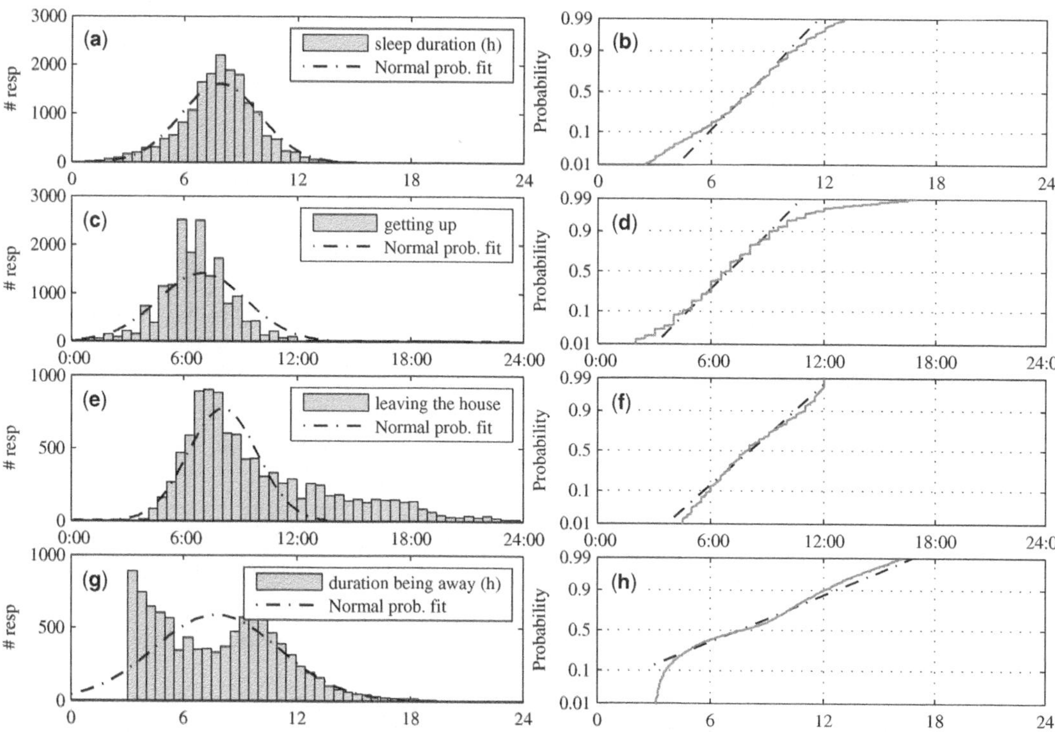

Figure 5.1 Frequency distribution and cumulative frequency distribution for (a, b) the duration of sleep, (c, d) time of getting up, (e, f) time of leaving the house and (g, h) duration of being away from home (for >3 hours) for all respondents of the 1995 survey (U.S. Department of Labor; Bureau of Labor Statistics)

The measurements are converted to SERPs (Buchberger and Wells 1996). In this chapter the simulation results from the PRPSym and SIMDEUM are compared to the SERPs of the measurements of 31 days in May and June 2007.

Table 5.5 Statistics for the diurnal patterns of different types of people for weekdays

		Weekday					
		Child	Teen	Adult with out-of-home job	Adult without out-of-home job	Senior	Total
Time of getting up	μ	8:00	8:00	7:00	7:00	7:00	7:00
	σ	2:30	2:30	2:30	2:00	2:00	2:30
Time of leaving the house	μ	8:15	8:15	8:00	13:00	13:00	8:00
	σ	1:45	1:45	1:45	3:00	3:00	1:45
Duration of being away	μ	7.25 h	7.25 h	8.0 h	6.0 h	6.0 h	7.75 h
	σ	2.8 h	2.8 h	3.4 h	3.0 h	3.0 h	3.3 h
Duration of sleep	μ	9.0 h	9.0 h	8.0 h	8.0 h	8.0 h	8.0 h
	σ	2.0 h	2.0 h	2.0 h	2.0 h	2.0 h	2.0 h

5.2.4 Comparing the two models

In the development and use of both models, the following steps can be discerned: Data collection; data analysis; model building; defining input data; running model; obtaining output data. The two models are being compared on several of those aspects.

When the actual models are considered as white boxes, it is interesting to compare the simulation steps during the running of the model (Section 5.3.1).

When the actual models are considered as black box models, the user of a model is only interested in the input data: how much effort is needed in collecting the input data for a location other than Milford, OH? A comparison on practical issues is considered for that purpose (Section 5.3.2).

The input parameters differ. However, some of the SIMDEUM input parameters can be translated to PRP input parameters and they can thus be compared. A qualitative comparison is done on the intensity and duration of pulses (Section 5.3.2).

The output of both models is a demand pattern for a single home. The measurements and the results of the two models are compared on two levels: a single home and the sum of 20 homes. For the single home evaluation (Section 5.3.4), 651 measurement patterns (21 homes, 31 days), 4000 PRP patterns and 4000 SIMDEUM patterns (40 homes, 100 days) were used. For the evaluation of the sum of 20 homes (Section 5.3.5), 50 patterns were constructed from measurements and model results. In Section 5.2.5, the parameters that were used for this comparison are explained.

5.2.5 Parameters to compare measurements and simulation results

The parameters from Table 5.6 were used to compare measurements and simulations (Chapter 3). The goodness-of-fit was determined with three parameters, viz. two measures for the error between empirical and modelled distributions (Mean Error, *ME*, and Root Mean Square Error, *RMSE*) and a measure for the similarity in shape between empirical and modelled distributions (coefficient of determination R^2).

The parameters Q_{max}, V, n_{pulse}, n_{hours}, Q_{CFD} and ΔQ_{CFD} are presented as cumulative frequency distributions. The goodness-of-fit for the cumulative frequency distributions is determined with ME, RMSE and R^2 evaluated at each integer percentile between 1 and 100 (1, 2, 3 to 100). Q_{day} is a pattern;

the goodness-of-fit is expressed as ME, RMSE and R^2 of Q_{day}. The ME and RMSE are presented as absolute values and as percentages of the means of the measurements. For Q_{CFD} and ΔQ_{CFD} the relative values are not meaningful because the averages of the measurements are close to 0. When ME and RMSE $\leq 30\%$ and $R^2 > 0.7$, the model fit is considered to be good.

Table 5.6 Overview of parameters used for validating simulation results

Parameter	Description	Temporal scale	Spatial scale
Q_{max}	Maximum flow per day	1 s, 1 min, 1 h	Single home, Σ 20 homes
V	Total water use per day	1 day	Single home, Σ 20 homes
n_{pulse}	Number of pulses per day	1 s	Single home, Σ 20 homes
n_{hours}	Number of hours of water use / busy hours	1 h	Single home, Σ 20 homes
Q_{day}	The water demand over the day	10 min, 1 h	Σ 20 homes
ρ	Correlation coefficient (the average and standard deviation of ρ are determined for a set of 50 measured or simulated patterns)	1–60 min	Single home, Σ 20 homes
Q_{CFD}	Cumulative frequency distribution of water demand over the day (CFD of flows)	1 s, 1 min, 15 min	Single home, Σ 20 homes
ΔQ_{CFD}	Cumulative frequency distribution of change in water demand over the day (CFD of flow accelerations).		

5.3 RESULTS: COMPARING THE TWO MODELS

5.3.1 Comparing the two models on underlying principles

The two models were compared first of all on their underlying principles, approaching the models as white boxes. This comparison shows the similarities and differences in modelling approaches and mainly focuses on the development of the model, rather than the use of the model. The latter is discussed in the next section.

The basics of the PRP model and SIMDEUM are Eq. 5.1. Both models generate pulses with a specific intensity, duration and arrival time, and the sum of all pulses is a demand pattern of a single home.

The differences are shown in Table 5.7. In the PRP model the frequency of residential water use follows a Poisson arrival process with a time dependent rate parameter; the probability distribution functions for the intensity and duration are lognormal probabilities with equal parameters for all pulses.

SIMDEUM first determines the users and fixtures (end-use types) in a home, with their specific probability distribution functions for the intensity, duration and frequency of use and a given probability of use over the day. SIMDEUM then determines for each end use the frequency, intensity and duration from the end-use specific probability distribution functions. Then, the time of occurrence is determined from the user specific diurnal patterns.

Table 5.7 Comparison of the models as white boxes

<table>
<tr><th></th><th>Model building</th><th>Model running</th></tr>
<tr><td rowspan="2">PRP model</td><td>Arrival time
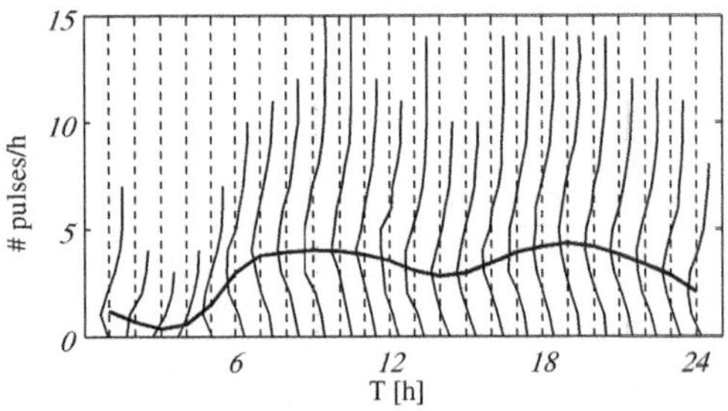
</td><td>Determine arrival times of pulses.</td></tr>
<tr><td>Intensity and Duration
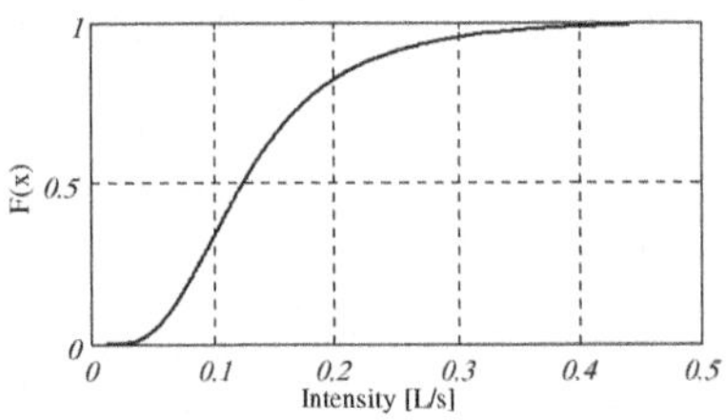

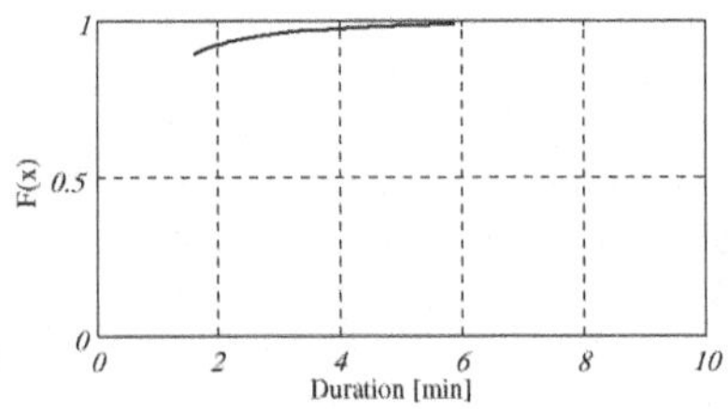
</td><td>Draw random numbers from the lognormal distributions to determine intensity and duration of pulses.</td></tr>
<tr><td>SIMDEUM</td><td>Users
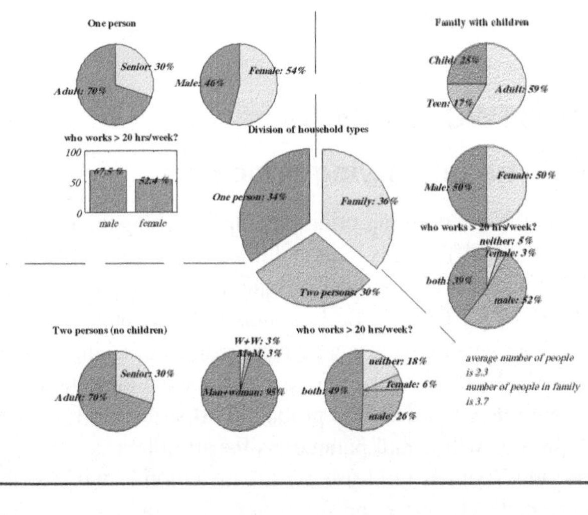
</td><td>Determine household type and number of users.
Determine ages and gender of users.
Determine sleep-wake times of users.

Add all pulses.</td></tr>
</table>

(*Continued*)

Table 5.7 *(Continued)*

Model building	Model running
Possession of appliances 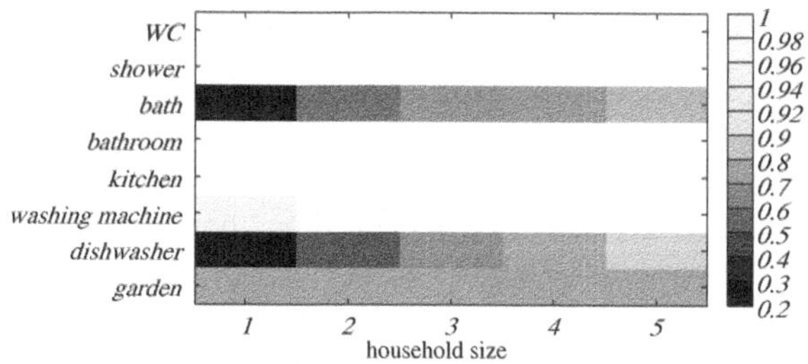	Determine for all types of end uses whether they <u>are present</u>.
Frequency 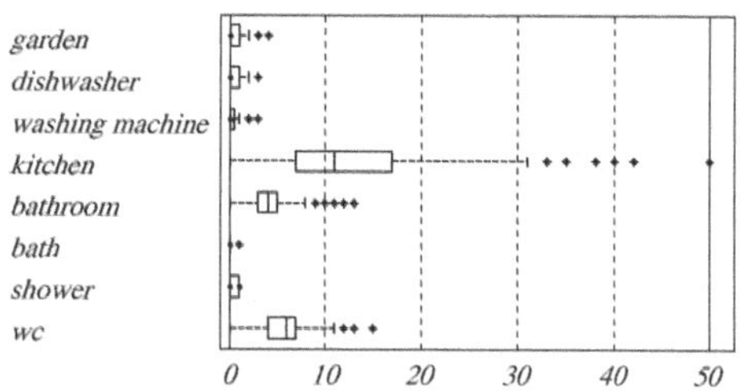	Determine, for all present end uses and users, the <u>frequency</u> of use from the applicable probability distributions (often Poisson).
Intensity and Duration 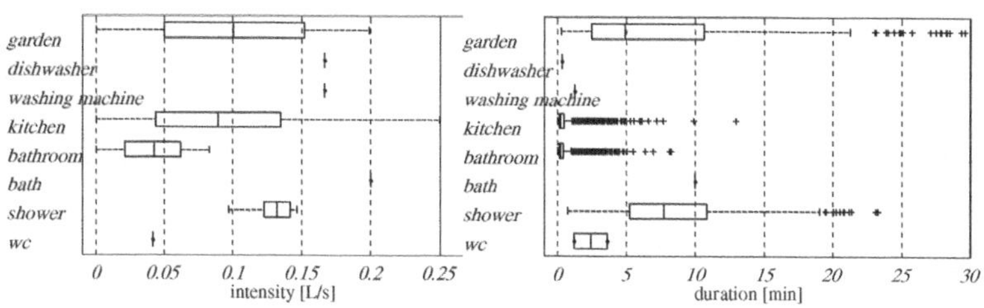	Determine for all uses the <u>intensity</u> and <u>duration</u> from the applicable probability distributions (often uniform for intensity and lognormal for duration).
Time over the day 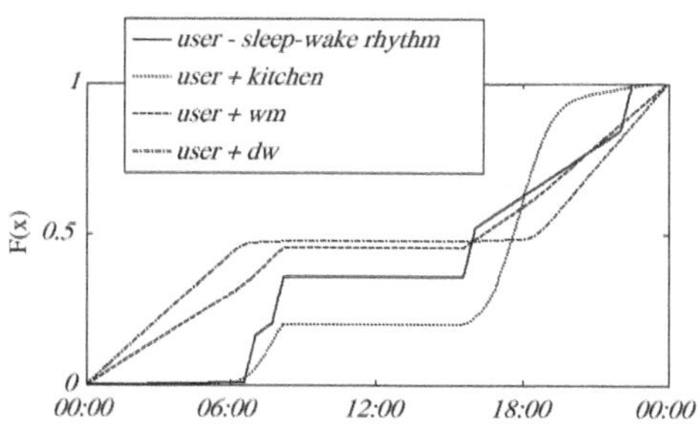	Determine for all uses <u>when during the day</u> it occurs from users' sleep-wake rhythm and preference hours for washing machine, dish-washer and kitchen tap. Adjust probability of use over the day to avoid coinciding uses of fixture. Add all pulses.

5.3.2 Comparing the black box models

A user who considers the models as a black box is interested in the input and output data only. Table 5.8 summarises the comparison of the two models.

Table 5.8 Comparison of the models as black boxes

	Data collection	Data analysis/Model input	Model results
PRP model	Measuring water demand at water meter 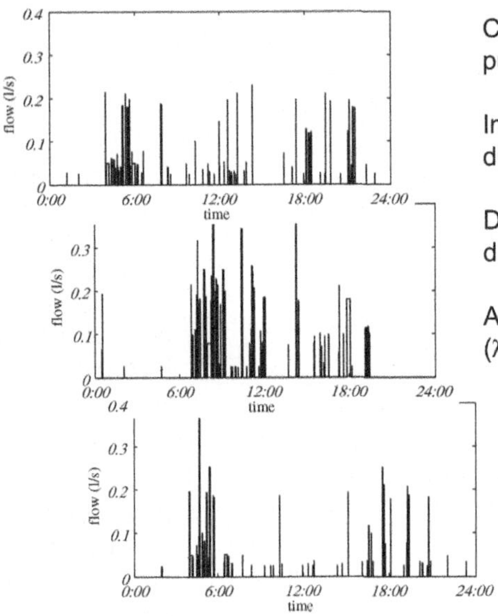	Convert measured data to pulses and analyse pulses Intensity – lognormal distribution (α, β) Duration – lognormal distribution (τ, ω) Arrival time – peak factors (λ, C_d)	For 1 home: 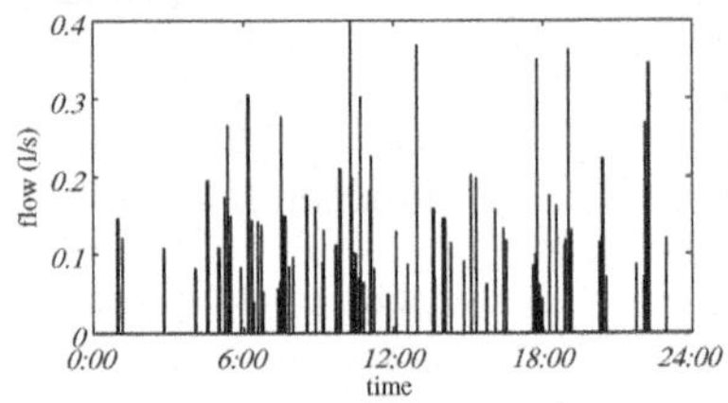
SIMDEUM	External data sources: (national / regional) population statistics (national) time use survey Water use survey Technical data of appliances	Household size – % one person, % two person, % family, average number of people, age distribution Possession of appliances Intensity, duration and frequency of use – average values for 8 fixtures: WC, shower, bath, kitchen tap, dishwasher, washing machine, bath, bathroom tap, outside tap Probability of use over the day – time of doing the dishes; time of getting up, leaving the house, returning home, going to bed	At individual taps: 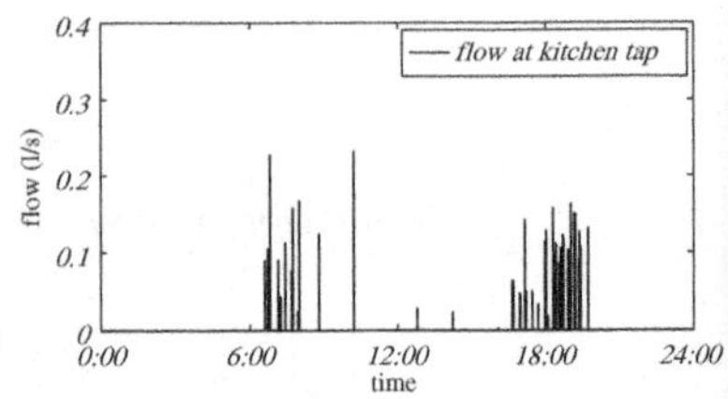 For 1 home: 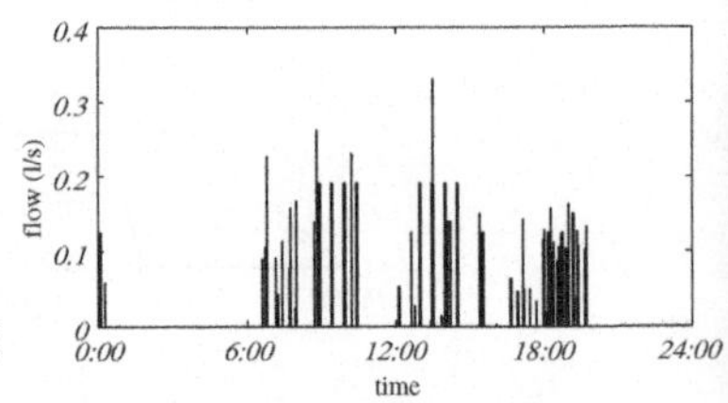

For the PRP model, water demand at the water meter level needs to be measured; the measured data needs to be converted into SERPs. The pulses are then analysed and the intensity, duration and arrival times are estimated. The output is a demand pattern of a single home.

For the end-use model SIMDEUM, the data can be found in the literature. For the Milford case, specific population statistics, time use survey data and technical appliance data were available. Part of the water use survey data could also be found for this specific case; the gaps were filled with Dutch data. The output is a demand pattern for each fixture; the sum of those demand patterns is a demand pattern of a single home.

The total number of parameters used in the PRP model is 6. The total number of parameters used in SIMDEUM is 99. For the end uses, a minimum of 24 parameters is required (Table 5.9), the Dutch case of Chapter 3 had 87 parameters and there were 30 Milford-specific parameters (excluding Dutch reuse of τ). The number of parameters describing the household composition is 21 (Table 5.4), or only 4 if no gender and age groups are discerned. The number of parameters describing the time use is 48 (count parameters in Table 5.5), or a minimum of 8 when no age groups are discerned.

Table 5.9 Number of parameters, describing the end uses. Grayed out are the parameters that are not counted in the total. The frequency of use (*F*) can be related to household size and age; the duration (*D*) can depend on the subtype of the end-use

End use	Minimum scenario						the Netherlands						Milford					
	P	*F*	*D*	*I*	τ	Total	*P*	*F*	*D*	*I*	τ	Total	*P*	*F*	*D*	*I*	τ	Total
Bath	1	1	1	1	0	3	5	5	1	1	0	12	1	5	1	1	0	7
Bathroom tap	1	1	1	1	0	3	1	1	2	1	0	4	1	1	2	1	0	4
Dishwasher	1	1	1	1	0	3	5	5	1	1	1	13	1	1	1	1	1	3
Kitchen tap	1	1	1	1	0	3	1	5	4	1	1	11	1	1	1	1	1	3
Outside tap	1	1	1	1	0	3	1	1	2	1	0	4	0	0	0	0	0	0
Shower	1	1	1	1	0	3	1	5	8	1	0	14	1	1	5	1	0	7
Washing machine	1	1	1	1	0	3	5	5	1	1	1	13	1	1	1	1	1	3
WC	1	1	1	1	0	3	1	10	5	1	0	16	1	1	1	1	0	3
Total						24						87						30

5.3.3 Comparing the input data

The intensity of the pulses in the PRP model is described by a lognormal distribution and the parameters α and β^2 are the mean and the variance of the pulse intensity, respectively (Table 5.1). The intensity of the pulses in the end-use model SIMDEUM is described by either fixed values or a uniform distribution (Table 5.3). When taking the frequency of use into account, the intensity of all pulses can be represented as a cumulative frequency distribution (Figure 5.2). The intensity of the pulses in the PRP model is higher than in SIMDEUM.

The duration of the pulses in the PRP model is described by a lognormal distribution and the parameters τ and ω^2 are the mean and the variance of the pulse duration (Table 5.1). The duration of the pulses in the end-use model SIMDEUM is described by either fixed values or a lognormal distribution (Table 5.3). When taking the frequency of use into account, the duration of all pulses can be represented as a cumulative frequency distribution (Figure 5.3). The duration of the pulses in the PRP model is shorter than in SIMDEUM. The retrieved PRP parameters led to mainly short pulses of 5 minutes or less. This means that showering, for example (circa 5 to 15 minutes), is almost never simulated as a single coherent event.

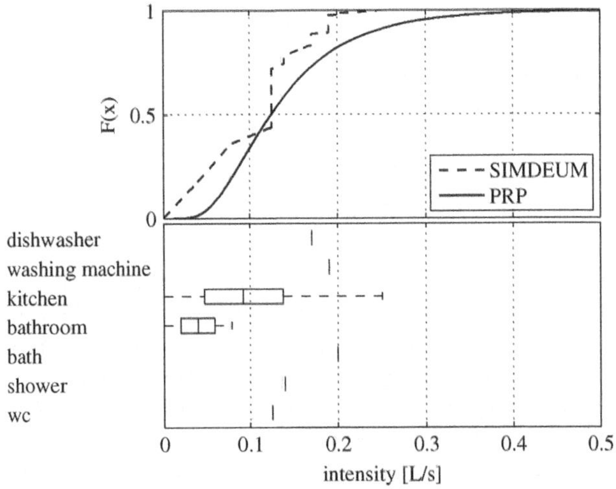

Figure 5.2 Intensity of pulses in PRP model and end-use model SIMDEUM; the lower part shows I per fixture

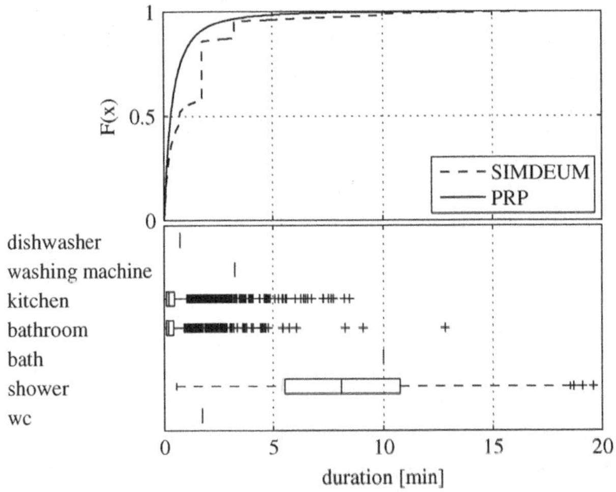

Figure 5.3 Duration of pulses in PRP model and end-use model SIMDEUM; the lower part shows *D* per fixture

5.3.4 Comparing the output data: results for single homes

For the single home, both the PRP model and SIMDEUM compare reasonably well to the measurements; the results are shown in Figure 5.4 and Figure 5.5 and in Table 5.10. On Q_{max} per second and per minute, SIMDEUM performs better than PRP; on Q_{CDF} and ΔQ_{CDF} (per second) both models perform well. On Q_{max} per hour and V per day, both PRP and SIMDEUM perform well.

On the number of pulses, neither model performs well. The reason may be that during the measurements some pulses that are under the detection limit may have been interpreted as multiple pulses. This leads to an overestimate of the number of pulses in the measured data. An over-occurrence of pulses may also have been caused by the conversion of measurements into SERPs. On *busy hours* SIMDEUM performs better than PRP; PRP has much fewer hours of zero water use.

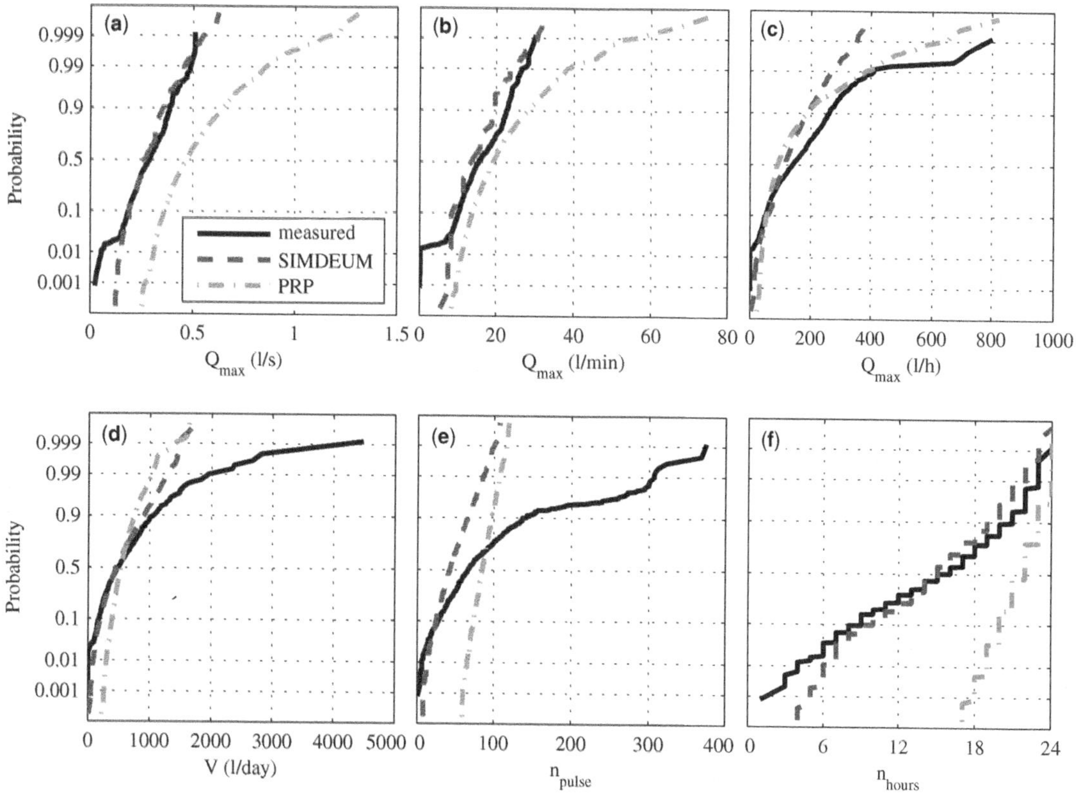

Figure 5.4 Comparing measurements and simulation results for single homes a) Q_{max} (L/s), b) Q_{max} (L/min), c) Q_{max} (L/h), d) V (L/day), e) n_{pulse}, f) n_{hours}

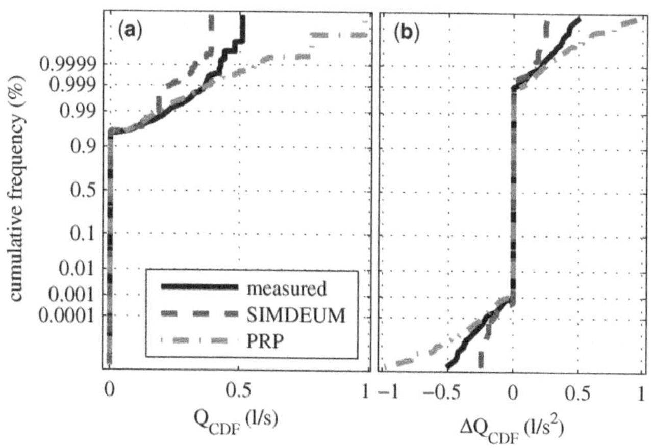

Figure 5.5 Comparing measurements and simulation results for single homes a) Q_{CFD} (L/s), b) ΔQ_{CFD} (L/s^2)

Table 5.10 Comparing measurements and simulation results for single homes at *ME*, *RMSE* and R^2 for Q_{max} (L/s, L/min, L/h), *V* (L/day), n_{pulse}, and n_{hours} of Figure 5.4 and Q_{CFD} and ΔQ_{CFD} of Figure 5.5

		SIMDEUM		PRP			SIMDEUM		PRP	
ME	Q_{max}	−0.0	−4.1%	0.21	70.5%	Q_{max}	−21	12.7%	4.5	−27.9%
RMSE	(L/s)	0.0	9.0%	0.22	75.1%	(L/min)	2.6	15.9%	6.4	39.2%
R^2		0.90		x			0.75		x	
ME	Q_{max}	−33.4	−21.3%	−39.1	−24.9%	*V*	−80.1	−13.2%	−37.5	−6.2%
RMSE	(L/h)	60.9	38.8%	50.6	32.2%	(L/day)	296.8	49.0%	360.6	59.5%
R^2		0.66		0.76			0.69		0.54	
ME	n_{pulse}	−38.7	−46.5%	2.45	2.9%	n_{hours}	−0.27	−1.8%	6.54	43.4%
RMSE		61.9	74.4%	55.1	66.2%		1.17	7.8%	7.41	59.1%
R^2		0.05		0.25			0.93		x	
ME	Q_{CFD}	0.0		0.0	<1%	ΔQ_{CFD}	0.0		0.0	
RMSE	(L/s)	0.0		0.1	<1%	(L/s^2)	0.0		0.1	
R^2		0.95		0.34			0.76		0.13	

The correlation coefficients for several time steps (Figures 5.6a, b, c) are predicted slightly better by SIMDEUM; PRP underestimates the correlation at the single home level.

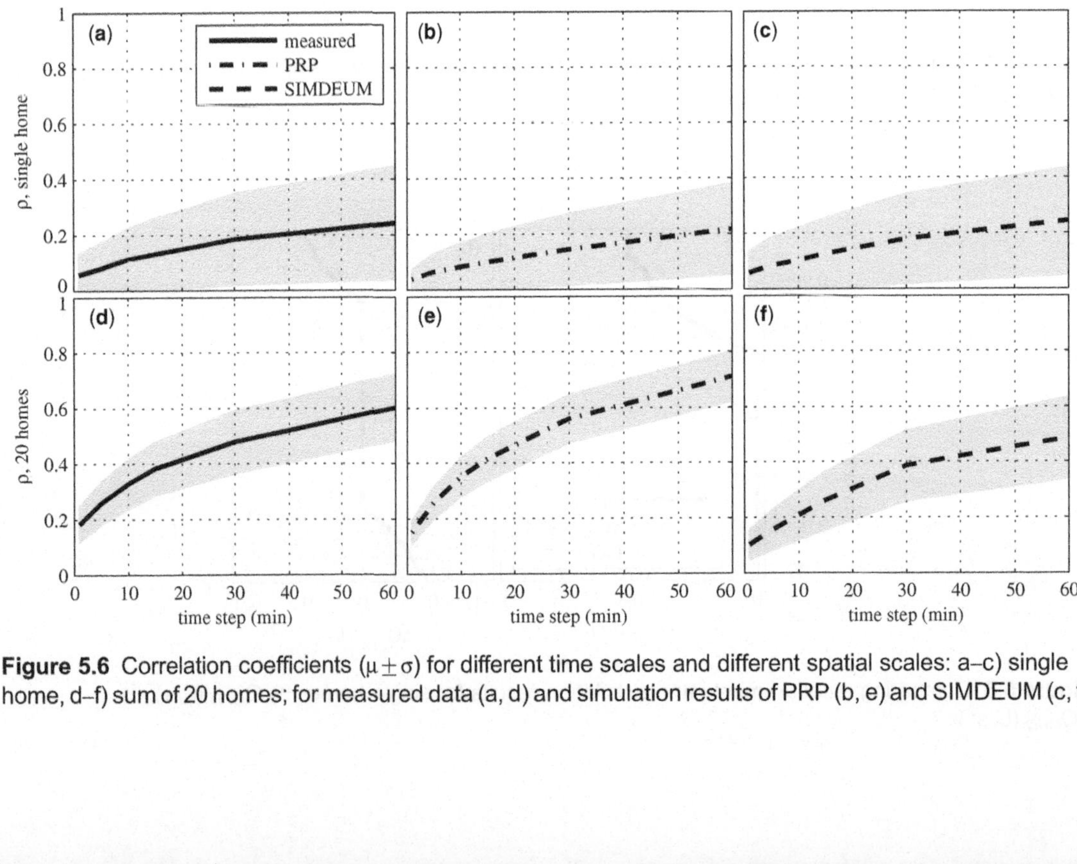

Figure 5.6 Correlation coefficients (μ ± σ) for different time scales and different spatial scales: a–c) single home, d–f) sum of 20 homes; for measured data (a, d) and simulation results of PRP (b, e) and SIMDEUM (c, f)

5.3.5 Comparing the output data: results for sum of 20 homes

For the sum of 20 homes both the PRP model and SIMDEUM compare reasonably well to the measurements; the results are shown in Figure 5.7 and Figure 5.8 and in Table 5.11. On Q_{max} per second PRP and SIMDEUM do not perform very well; on Q_{CDF} and ΔQ_{CDF} (per second) both models perform well. On Q_{max} per minute and per hour, and on V per day, PRP performs better than SIMDEUM.

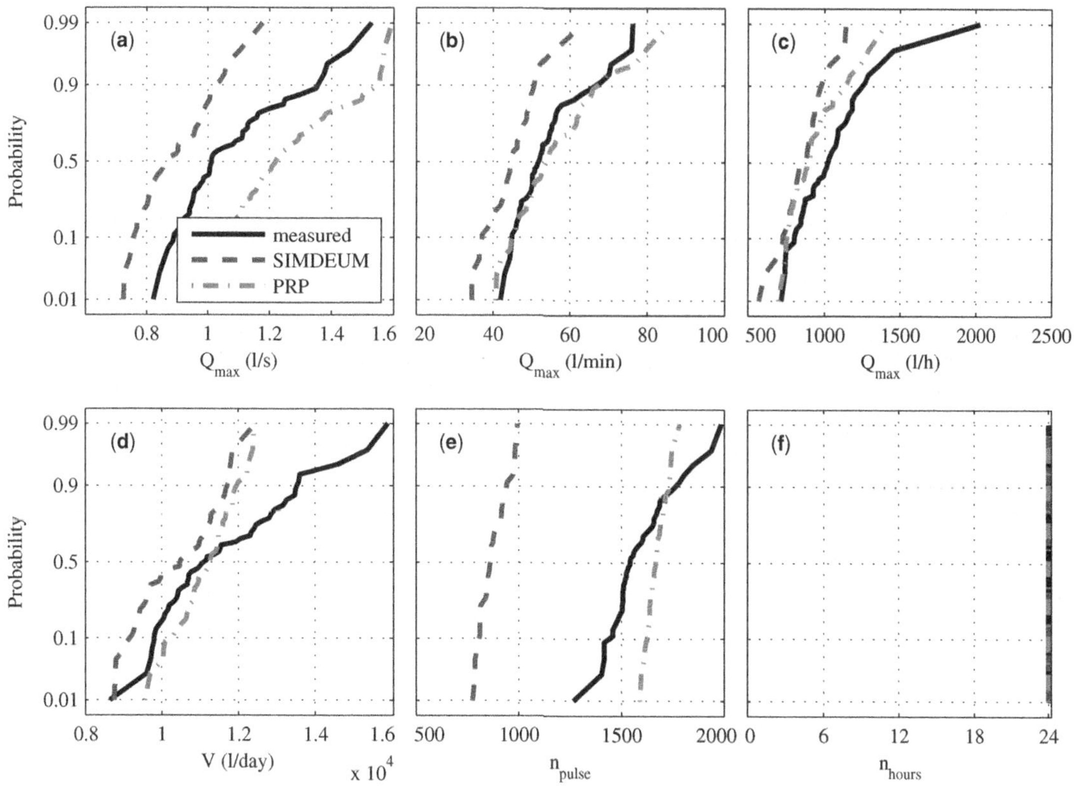

Figure 5.7 Comparing measurements and simulation results for sum of 20 homes a) Q_{max} (L/s), b) Q_{max} (L/min), c) Q_{max} (L/h), d) V (L/day), e) n_{pulse}, f) n_{hours}

On the number of pulses neither model performs very well; PRP does predict n_{pulse} better than SIMDEUM. On n_{hours} both PRP and SIMDEUM lead to demand patterns with 24 hours of water use.

The correlation coefficients (Figures 5.6d, e, f) for several time steps are overestimated by PRP and underestimated by SIMDEUM - with the absolute difference approximately equal.

On the pattern over the day (Q_{day}), PRP performs better than SIMDEUM (Figure 5.9). The mean and the variance of the simulated water demand patterns for both models are comparable to the mean and variance of the measured water demand patterns. SIMDEUM overestimates the night use and underestimates the daytime use.

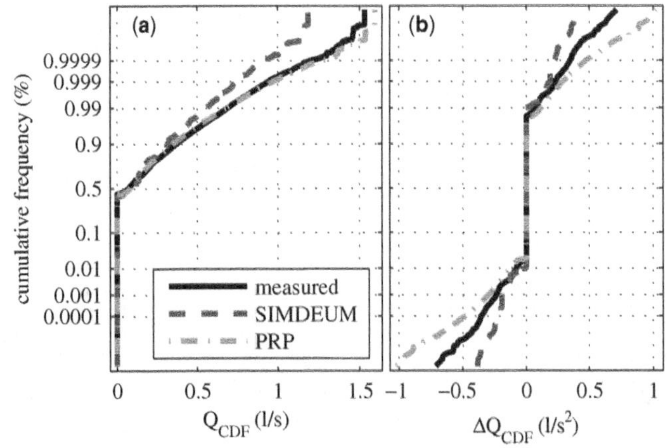

Figure 5.8 Comparing measurements and simulation results for sum of 20 homes a) Q_{CFD} (L/s), b) ΔQ_{CFD} (L/s^2)

Table 5.11 Comparing measurements and simulation results for sum of 20 homes at *ME*, *RMSE* and R^2 for Q_{max} (L/s, L/min, L/h), *V* (L/day), n_{pulse}, and n_{hours} of Figure 5.7, Q_{CFD} and ΔQ_{CFD} of Figure 5.8, and Q_{day} and C_d of Figure 5.9

		SIMDEUM		PRP			SIMDEUM		PRP	
ME	Q_{max}	−0.2	16.9%	0.2	17.2%	Q_{max}	−8.5	−15.8%	2.2	4.1%
RMSE	(L/s)	0.2	18.1%	0.2	17.5%	(L/min)	9.2	17.2%	3.3	6.2%
R^2		x		x			x		0.83	
ME	Q_{max}	−179.2	−17.2%	−124.0	−11.9%	*V*	−1.1e3	−9.6%	−0.4e3	−3.1%
RMSE	(L/h)	221.9	21.3%	152.8	14.7%	(L/day)	1.3e3	11.4%	1.1e3	9.2%
R^2		0.03		0.54			0.31		0.56	
ME	n_{pulse}	−713.3	−45.1%	87.9	5.6%	n_{hours}	0	0%	0	0%
RMSE		721.8	45.6%	130.5	8.3%		0	0%	0	0%
R^2		x		0.08			1.00		1.00	
ME	Q_{CFD}	0.0		0.0		ΔQ_{CFD}	0.0		0.0	
RMSE	(L/s)	0.1		0.0		(L/s^2)	0.0		0.0	
R^2		0.94		1.0			0.82		0.81	
ME	Q_{day}	−0.1	−9.5%	−0.01	−2.9%					
RMSE	(m^3/h)	0.19	40.1%	0.10	21.9%					
R^2		0.34		0.80						

5.4 DISCUSSION

5.4.1 Practicality

The PRP model has only a few parameters and, therefore, this is a relatively straightforward model. Owing to strong similarities among water-use fixtures and household appliances in homes across

North America, the patterns and properties of indoor water use are fairly consistent from coast to coast (Mayer et al. 1999; Solley et al. 1998). Hence, the values listed in Table 5.1 provide reasonable initial estimates of the key PRP parameters for indoor residential water use. For other countries, new measurements are required to establish the PRP parameters. Obtaining the PRP parameters requires many (expensive) measurements. For example, the parameters of Milford, Ohio (Buchberger et al. 2003), were obtained from 30 days of measurements of 21 homes on a per second basis. These measurements were required to prove the PRP concept. While future field studies would not involve such an exhaustive campaign, there is, nonetheless, a large effort and high costs. Spectral analysis of residential water demand time series (Alcocer-Yamanaka and Tzatchkov 2009) showed that, in order to construct a PRP or NSRP model, a maximum measurement interval of 200 s is required, i.e. measurements with short time intervals are required.

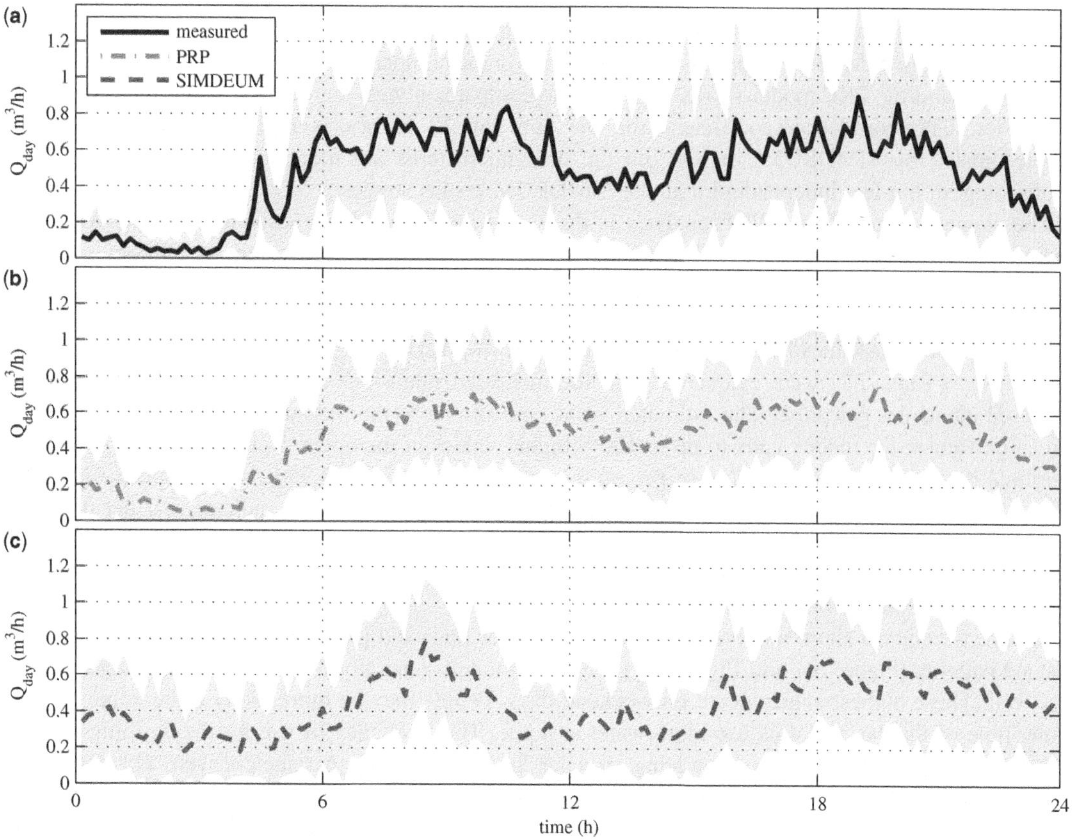

Figure 5.9 Comparing measurements and simulation results for sum of 20 homes on mean and variance ($\mu \pm \sigma$) of Q_{day} (m^3/h) at a time scale of 10 minutes for a) measured data; b) simulation results of PRP; c) simulation results of SIMDEUM

SIMDEUM has many parameters and, therefore, it is not a straightforward model to build. However, the end-use model is intuitive, because all parameters have a physical meaning. The input data for SIMDEUM were found in publicly available databases and research papers. It took approximately one day to construct the total model for the 21 Milford homes; this means SIMDEUM is a relatively easy model to run. The distribution functions that were found from the Dutch data analysis, such as a Poisson distribution for the number of toilet flushings, probably can be generally applied. The average values of frequency of use are most likely different for different countries and different cultures. For example, the Dutch take showers and use their baths seldom; the English take a bath much more often. Consequently, only average values are required for the input of SIMDEUM, the statistical distributions can be re-used. These average values are typically available in the literature.

5.4.2 Performance

An issue is that it is difficult to determine how well the PRP model performs compared to the measurements, since the simulation parameters were derived from the same measurements. Both models compare well with the measurements. SIMDEUM appears to work better on the comparison with the demand patterns of the single home and the PRP model works better on the comparison with the demand patterns of the sum of 20 homes. The correlation coefficient for the sum of 20 homes is overestimated by PRP and underestimated by SIMDEUM. In real life the demand patterns are more independent than the PRP results suggest, but less independent than the SIMDEUM results suggest.

The elemental unit for the PRP analysis is the single-family residence (Buchberger and Wu 1995). Therefore, it was expected that the PRP model would perform better for the sum of 20 homes than for the single home. Q_{day} of 20 homes is very well described by the PRP model; the measurements show a larger variation than the model does.

The PRP model assumes larger pulse intensities than SIMDEUM (Figure 5.2). The PRP parameters for the intensity (Table 5.1) were estimated from the measured intensities at all time steps, instead of the measured intensities per pulse. The estimated values are similar to the measured values for the lowest 95 percentile values; the model leads to up to 100% higher values in the upper 5 percentile compared to the measured data. The mean and variance per pulse are 5.86 L/min and 15.95 L/min^2 respectively. When these values are used for α and β^2, Eq. 5.4 is not valid. These lower values indicate that per pulse, the PRP model overestimates the intensity. This leads to the PRP model overestimating the maximum flow per second of a single home (Figure 5.4a). The approach of simulating the end uses with their specific flow rates (intensities) works much better in this case.

The measurements show relatively high water uses per hour and per day for the top 1–10% of water demand patterns (Figure 5.4 and Figure 5.7). The two models generally do not predict this highest value very well. These highest values occur on weekend days. People may be more likely to be home during a longer time of the day and thus use more water at home. The frequency of use for the washing machine may be higher; the duration of showering may be longer. The PRP parameters were not determined separately for working days and weekend days. As there are fewer weekend days in a week, the PRP model is more likely to fit the working days. However, a different set of parameters is probably applicable for weekend days. This may improve the fit for the highest water use values. To properly simulate weekends with SIMDEUM, the frequency of use and the sleep-wake rhythms of the users need to be adjusted. It is unclear what would be proper values for weekends.

When SIMDEUM was run with the Dutch time-use data, the SIMDEUM pattern showed a Dutch water demand pattern with low night use, and a distinctive (and higher) peak in the morning (Blokker et al. 2008). By using the American time-use data (ATUS), the shape of the demand over the day resembles the Milford measured demand pattern much more closely. SIMDEUM overestimates the night use and underestimates the daytime use and underestimates the cross-correlation. The explanation may be that the ATUS data for the entire USA are more variable than they would be for the 21 Milford homes. The Milford residents in 1997 may have had fewer night time jobs than the average American resident.

5.4.3 Application

The PRP model describes the measured flows very well. From the analytical description that the PRP model provides, a lot of mathematical deductions can be made. Thus, the PRP model is a descriptive model with the potential to provide insight into some basic elements of water use, such as peak demands (Buchberger et al. 2008), travel times (Buchberger et al. 2003) and cross-correlation (Li et al. 2007).

SIMDEUM goes one layer deeper than the PRP model, namely to the fixture level instead of the household connection level and therefore uses more parameters (99 in the Milford case). SIMDEUM is a Monte Carlo type simulation that can be used as a predictive model. SIMDEUM can be applied in scenario studies to show the result of changes in water-using appliances and human behaviour. Also, SIMDEUM goes one layer deeper than the PRP model and can thus give more insight into what happens inside the single home, e.g. for describing hot water flows or the water use of the kitchen specifically.

5.5 CONCLUSION

The results from both the PRP model and SIMDEUM fit the measured flow data very well. The PRP model requires flow measurements and accordingly represents the measured data well. The end-use model SIMDEUM requires sociologic data and can thus be tuned to the region under study. The required SIMDEUM input data for the Milford study could easily be collected. Therefore, the transferability of SIMDEUM is demonstrated.

SIMDEUM appears to work better on the comparison with the demand patterns of the single home and the PRP model works better on the comparison with the demand patterns of the sum of 20 homes. The PRP model is a descriptive model, whereas SIMDEUM is more of a predictive model. Accordingly, the two models have different areas of application.

Chapter 6

A bottom-up approach of stochastic demand allocation in hydraulic and water quality modelling*

ABSTRACT

An "all pipes" hydraulic model of a drinking water distribution system was constructed with two types of demand allocations. One is constructed with the conventional top-down approach, i.e. a demand multiplier pattern from the pumping station is allocated to all demand nodes with a correction factor to account for the average water demand on that node. The other is constructed with a bottom-up approach of demand allocation, i.e. each individual home is represented by one demand node with its own stochastic water demand pattern. The stochastic water demand patterns were constructed with the end-use model SIMDEUM on a per second basis and per individual home. This was done for two drinking water distribution systems: one small system, approximately 150 homes, and one larger system, ca. 1000 homes.

Both systems were tested in a real life situation. The flow entering the test area was measured and a tracer test with sodium chloride was performed to determine travel times. The two models were validated on the total sum of demands and on travel times. In the small area a sensitivity test with respect to the resulting residence times was performed for several model parameters: time step, spatial aggregation, spatial correlation, demand pattern and number of simulation runs.

The studies showed that the bottom-up approach leads to realistic water demand patterns and travel times, without the need for any flow measurements or calibration. The stochastic approach of hydraulic modelling, with a 15 minute time step, some spatial aggregation and 10 simulation runs, gives insight into the variability of travel times as an added feature beyond the conventional way of modelling.

6.1 INTRODUCTION

The goal of drinking water companies is to supply their customers with good quality drinking water 24 hours per day. With respect to water quality, the focus has for many years been on drinking water treatment. Recently, interest in the water quality of a drinking water distribution system (DWDS) has been growing. Water age is an important aspect of water quality in a DWDS as it influences disinfectant

---

*Parts of this chapter were based on: Blokker, E. J. M., Beverloo, H., Vogelaar, A. J., Vreeburg, J. H. G., and van Dijk, J. C. (2011). "A bottom-up approach of stochastic demand allocation in a hydraulic network model; a sensitivity study of model parameters." *Journal of Hydroinformatics*, 13(4), 714–728, doi:10.2166/hydro.2011.067 with permission from the copyright holders, IWA; Blokker, E. J. M., Vreeburg, J. H. G., Beverloo, H., Klein Arfman, M., and van Dijk, J. C. (2010). "A bottom-up approach of stochastic demand allocation in water quality modelling." *Drink. Water Eng. Sci.*, 3(1), 43–51.

residual, disinfection by-products, nitrification, bacterial regrowth, corrosion, sedimentation, temperature, taste and odour (EPA 2002). More specifically, the maximum water age (or residence time) is most important (Machell et al. 2009).

The key element of a water quality model for a DWDS is a detailed hydraulic model (Slaats et al. 2003; Vreeburg 2010), which not only takes into account the maximum flows but also the flows at the preceding time steps (Powell et al. 2004; Slaats et al. 2003; Vreeburg and Boxall 2007). A hydraulic model with an accurate simulation of the occurrences of turbulent and laminar flows and stagnant water is required. Therefore, knowledge of the water demand on a more detailed level is essential (Chapter 2). The required detail in temporal and spatial scale is to be determined. This chapter investigates the required detail.

One way of improving a model's accuracy is by calibration. Calibration is usually done with pressure measurements by adjusting wall roughness coefficients and the status of valves, for which several optimization techniques are available (Kapelan 2002). For this calibration a discernible head loss is required; in the periphery of the drinking water distribution system, where velocities and head losses are low, calibration based on pressure is almost impossible. Jonkergouw et al. (2008) showed that calibration of demands can be done by using water quality measurements (in their case chlorine levels). They concluded that average daily demands can be determined with high precision, but that substantial measurement errors in the calibration data (i.e. water quality data) do not allow for an accurate calibration of the demand multiplier pattern (DMP) that construct the diurnal pattern. Pasha and Lansey (2010) have shown that water quality predictions of residual chlorine in a DWDS are very sensitive to uncertainty in demand and the bulk and wall reaction coefficients, and hardly sensitive to pipe diameter and wall roughness. Calibration of the diameter and wall roughness by means of pressure measurements may therefore not be required. Water quality measurements are preferably used for the calibration of reaction coefficients but not also for the calibration of demands.

Modelling water quality in the DWDS requires a different approach in demand allocation, where the demands show less auto- and cross-correlation and are determined on smaller temporal and spatial scales than the conventional 'top-down' approach of demand allocation (Chapter 2). Here, top-down demand allocation means that a DMP (e.g. measured at the pumping station) is allocated to the demand nodes with a base demand to account for the average water demand on that node, thereby applying strong spatially correlated water demand patterns among all nodes.

A different way is to use a 'bottom-up' approach of demand allocation. This means that unique stochastic water demand patterns are modelled for each individual home for each day of the week and a unique water demand pattern is constructed for each demand node by summation of the individual household's water demand patterns. In the traditional approach of top-down demand allocation, the cross-correlation is assumed to be equal to 1 and the auto-correlation is usually high when a time step of 15 min or 1 h is used. A cross-correlation of 1 results in a limited number of flow direction reversals in a network model. A high auto-correlation means that the flow over the day is relatively constant and the model will show no periods with stagnant water and possibly a limited period of turbulent flow. In case the actual flows are not strongly correlated, flow direction reversals and periods of stagnancy and turbulent flows will occur. A traditional approach in demand allocation may therefore underestimate maximum residence times and dispersion.

The hypothesis is that a bottom-up approach of demand allocation results in a model with realistic demands, which show more resemblance with real demands with respect to instantaneous peak values and diurnal variability, and therefore leads to realistic residence times. For this model, no further calibration of demand is required. The hypothesis is tested by comparing this bottom-up approach against the traditional top-down approach and measurement results of two tracer studies with a conservative compound. The bottom-up demand allocation was done with the use of the end-use model SIMDEUM (Chapter 3).

The first part of the study was done on a small area of only 144 homes in the town Benthuizen. In this test area, the Taylor dispersion was limited. The measurements clarified how to properly do a tracer test. The sensitivity analysis of the model showed what spatial and temporal scales were required as well as how many model runs were required. The model was validated with measured flow patterns and the residence times of three locations.

The second part of the study was done on a larger network of about 1000 homes and a few hotels and beach clubs in the town Zandvoort. In this test area, no additional measures were take to limit the dispersion. The tracer test was done according to the findings of the first test area. No sensitivity analysis of the model was performed; the recommendation on spatial and temporal scales and the number of model runs from the first test were used. The model was validated with measured flow patterns and the residence times of four locations.

6.2 METHODS AND MATERIALS – BENTHUIZEN AREA

6.2.1 Generic methodology

A small distribution network was selected as a test area. In this network, the total water demand was measured and a tracer study was performed to determine the residence time towards three locations in the network. The network was operated in two different ways, viz. in a looped and a branched layout, and a continuous flow of 400 L/h was extracted.

An "all pipes" hydraulic model was constructed with a bottom-up approach of demand allocation of individual and unique stochastic demand patterns. A second model was constructed with a conventional top-down approach of demand allocation with a common DMP. The model results were compared with respect to the measured flow and residence time. A sensitivity test was done for the models. Demand patterns with various time steps were applied; demand patterns were allocated at the household connection and aggregated on the modelled pipe ends; different sets of demand patterns were applied.

6.2.2 The network

The selected network is situated in the town Benthuizen in the west of the Netherlands (near The Hague). The network was built in the mid 1970s and consists of 580 m of Ø100 mm Asbestos Cement pipes, 380 m of Ø110 mm, Ø63 mm and Ø50 mm PVC pipes, and 70 m of Ø80 mm lined Cast Iron pipes and supplies 144 homes (Figure 6.1). The pipes have well-defined internal diameters and wall roughness. The roughness of AC and PVC pipes is not known to be age-related; for Cast Iron, there is an unidentified relation with age. For the 70 m of Cast Iron in this network, the wall roughness was assumed to be 0.5 mm.

The annual water use in the network was determined from the water meter readings of 2004 of the Dunea Water Company. On average, 314 L per home per day was registered. This was confirmed by flow measurements in 2004 on a district metered area (DMA) of ca. 1200 homes (Beuken et al. 2006) which encloses the network under study.

The supply area Vlietregio, in which Benthuizen is located, supplies about 16,000 (mainly residential) connections; its flow is continuously measured. The measured water demand patterns of this supply area for the period 21–30 July 2007 are indicated by DMP_{PS}, where PS stands for pumping station. The 2006 study showed that this network has no leaks. Low leakage is common in the Netherlands (Beuken et al. 2006; Geudens 2008). Since dispersion could have a large effect on water quality modelling (Li 2006), dispersion in the network was limited by applying an additional demand of 400 L/h. This measure ensured the absence of stagnant water and an average Reynolds number of 5500, i.e. a turbulent flow during most of the day.

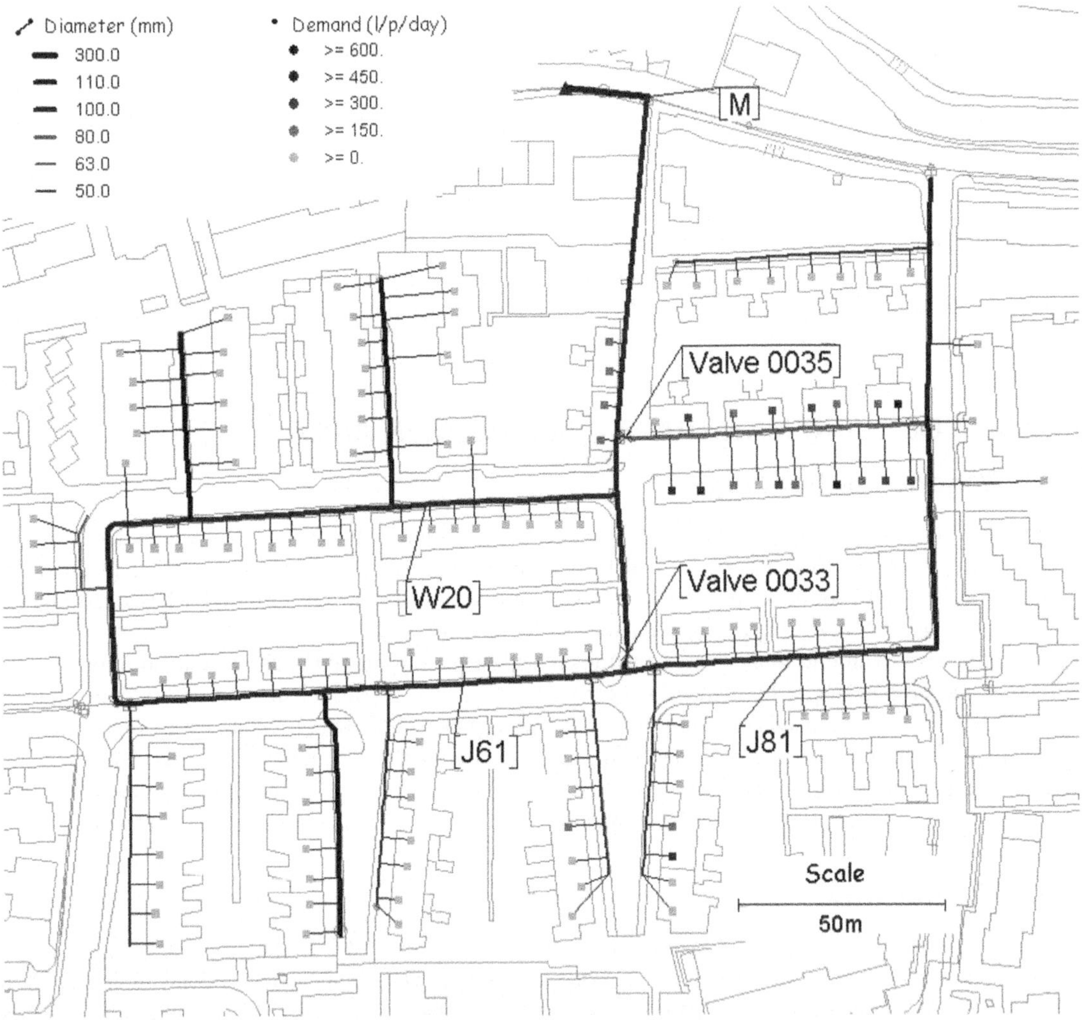

Figure 6.1 Network layout. Water enters at location 1. Valves 0033 and 0035 are closed in the branched layout and open in the looped layout. Valve 0035 is placed at the 80 mm CI main

The drinking water was distributed without any disinfectant, as is common in the Netherlands. A tracer study with NaCl was performed between 24 and 30 July 2007. Some valves were permanently closed during the measurement period to isolate the area from the rest of the DWDS. Two other valves (0033 and 0035) were operated to set the network layout to either a branched or a looped system. The valves were closed from Tuesday 24 to Thursday 26 July; they were open from Friday 27 to Sunday 29 July.

6.2.3 Measurement setup for the tracer study

Four measurement locations were selected (Figure 6.1). Location 1 is located at the entrance of the isolated test area. Locations 2, 3 and 4 are located on the central Ø100 mm AC main. Note that in the branched network layout, the water travels from location 1, to 2, to 3 and then to 4. In the looped network layout, this is not necessarily the case.

Sodium chloride (NaCl) was used as a tracer and the electrical conductivity was measured; from these measurements, the residence time was determined. NaCl has several advantages for use as a tracer, viz. at a well measurable dosage it causes no disruption or health risk to customers; it yields results of good accuracy and it is low cost (Skipworth et al. 2002). At location 1, NaCl was dosed to a fixed concentration in order to raise the electrical conductivity (EC, in mS/m) by a measurable amount: EC $\approx$ 44 mS/m without dosage, and EC $\approx$ 58 mS/m with dosage. The tracer was dosed in pulses of 4 hours on and 4 hours off. This means that, per day, 6 positive and 6 negative step inputs were induced.

In order to reach a fixed concentration, the flow was measured (Endress + Hauser Promay W) and the dosage was controlled. A static mixer ensured a constant concentration of the tracer over the pipe cross-section. To overcome the head loss through the static mixer and to establish a fixed head, a pump was placed at location 1. The instantaneous flow was logged every minute; this resulted in five full days of flow measurements at location 1 (3 weekdays, 2 weekend days). The measured flow patterns were converted into an average DMP for weekdays and one for weekend days, which is denoted DMP_{REF}.

At all 4 locations, the EC was measured (LIQUISYS M CLM223). The monitoring systems required a continuous 40 l/h extraction at the measurement locations. An extra 400 l/h was extracted at location 4, to ensure turbulent flows during most of the day.

The residence time between location 1 and 2, 3 and 4 was determined from the time between the centres of the ascending and descending tails (at ca. 51 mS/m) of the measured EC pulses. As laminar flows were mostly avoided, the pulses retained their shapes. Figure 6.2 shows that at 15:12, the residence time between locations 1 and 2 was equal to 0.5 h; at 18:37 the residence time between locations 1 and 2 was equal to 0.3 h. The residence time varies over the day and between days. This variation is considered in both the measurements and the hydraulic model.

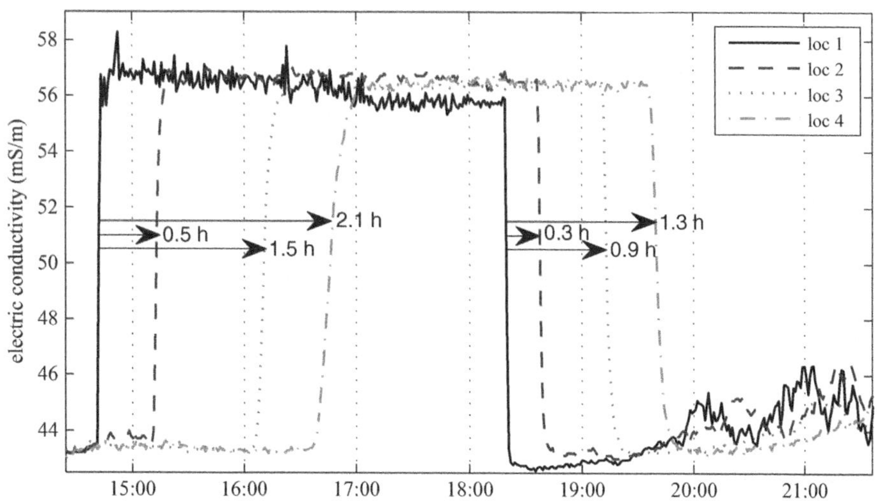

Figure 6.2 EC at the measurement locations (Tuesday, 24 July 2007) and the residence times between locations 1 and 2, 3, and 4

6.2.4 Hydraulic model and demand allocation

Wallingford's InfoWorks® was used as a hydraulic network model solver. Basically, two models were constructed that are distinguished by demand allocation. $Model_{TD}$ is the model with the top-down approach of demand allocation; $Model_{BU}$ is the model with the bottom-up approach of demand allocation.

Each of the 144 homes was defined as a customer point (the InfoWorks entity that comprises water demand) and these customer points were connected to nodes on the mains. The hydraulic model $Model_{TD}$ has consolidated demand nodes at pipe ends and junctions (the open circles in Figure 6.1), i.e. demands are not applied per home but per node. The hydraulic model $Model_{BU}$ has unique individual demand nodes for all homes; the demand nodes are located at the stop taps of the homes (depicted by the connecting lines between demand nodes and distribution mains in Figure 6.1).

Measurement locations 1, 2, 3 and 4 were assigned a continuous extraction of 40, 40, 40 and 400 l/h, respectively. The demand allocation at the customer points was done in two ways:

1. In $Model_{TD}$ an identical DMP was allocated to all customer points with a correction factor to account for the average demand per day. The utilised DMP are a) DMP_{PS}, b) DMP_{REF} and c) DMP_{SIM} (Table 6.1). The average demand for each customer point was assigned based on the water meter reading of 2004.
2. In $Model_{BU}$ a unique stochastic water demand pattern was assigned to each individual customer point (Section 6.2.5).

Table 6.1 Overview of Demand Multiplier Patterns in $Model_{TD}$

	DMP_{PS}	DMP_{REF}	DMP_{SIM}
name	DMP of pumping station	Reference DMP of test area	Simulated DMP of test area
origin	Supply area Vlietregio	Test area	$Model_{BU}$ of test area with SIMDEUM demand patterns
# homes	16,000	144	144
time	21–30 July 2007	24–29 July 2007	n.a.
# weekdays	6	3	10
# weekend days	4	2	10
original time step	5 min, average flows	1 min, instantaneous flows	1 s
remark		Continuous flows for monitoring systems were subtracted	

In accordance with the measurement period, the branched system was modelled with weekday patterns; the looped system was modelled with weekend patterns.

The water demand patterns of the individual homes were generated on a 1 s time base. The generated water demand patterns were time averaged over different time scales before assigning them to the individual customer points in the hydraulic model. The hydraulic model was run with a hydraulic time step equal to the pattern time step, which was set to 1 min, 15 min and 1 h in different computer runs.

The model was not calibrated on pressure; the selected network has well-known pipe lengths, diameters and wall roughness, and a fixed head at the entry point.

6.2.5 Water demand pattern generation

The end-use model SIMDEUM (Chapter 3) was used as a water demand pattern generator. SIMDEUM input consists of information on the number and age of residents, the residents' sleep-wake rhythm and possession of and behaviour with respect to water-using appliances. Generic Dutch data was used for the water-using appliances (Kanne 2005a) and time use (SCP 1995). Specific census data for the town Benthuizen was used, this information is available in the Netherlands per postal code area (CBS; Table 6.2). This means that only the data of Table 6.2 was specific, all other input data for SIMDEUM was found from the average Dutch data (Chapter 3).

Table 6.2 Specific Benthuizen input data into SIMDEUM; data of postal code 2731 in 2006 (CBS)

		Benthuizen
House-holds	One person households	22%
	Households without children	30%
	Households with children	48%
	Average household size	2.8
Age distribution	0 to 15 years old	23%
	15 to 25 years old	14%
	25 to 45 years old	26%
	45 to 65 years old	27%
	65 years and older	11%

For each of the 144 homes, 20 unique water demand patterns on a time scale of 1 s were generated; 10 patterns for weekdays and 10 patterns for weekend days. The patterns were then temporally aggregated (over 1 min, 15 min and 1 h) and divided by the average daily demand as obtained from the water meter readings. This led to 10 DMPs for each home. These DMPs do not necessarily have an average of 1 because they were divided by the average daily demand and not the average of the specific simulated demand. It was tested if 10 runs were enough to get a good view of mean and standard deviation of the residence times (see Sec. 6.2.7).

A table was constructed which cross-references customer point identification, average daily demand (L/day) and the demand category identification number (ID). This table was imported into the InfoWorks model. One demand category ID is linked to each of the 144 homes and remains the same for all 10 different patterns. The demand multiplier patterns were exported into a text file of a specific InfoWorks format - a so-called '.ddg' file. For each 'demand category', information on the demand category ID, the number of multipliers, the time step and multiplier per time step are written to the file. This led to 20 computer-generated.ddg files containing 144 DMPs per time scale. For the three different time scales (1 min, 15 min and 1 h) this means that 60.ddg files were created. These.ddg files were imported for the different model scenarios.

6.2.6 Water quality model

The hydraulic model was run with the water quality option enabled. This allowed for the determination of the residence time, which InfoWorks calls "water age". To determine the residence time the simulation run was set to 48 hours, where the diurnal demand patterns were repeated at hour 24 to 48. Because the residence times in this network do not exceed 24 hours, the diurnal patterns of the second day were not altered. One model run took less than one minute.

6.2.7 Sensitivity analysis and model validation

Twelve different scenarios were modelled (Table 6.3), which are related to network layout (branched and looped), different hydraulic time steps (1 min, 15 min and 1 h) and different demands. Scenarios 1, 3 and 5 were analysed to determine the influence of the temporal scale. Scenarios 7, 9 and 11, and 8, 10 and 12 were studied to determine the influence of DMP. Scenarios 3 and 9, and 4 and 10 were examined to determine the influence of the top-down approach and the bottom-up approach of demand allocation.

Table 6.3 Model scenarios

Scenario	Hydraulic model	Layout	DMP	Specifics	Time step	# runs
1	Model_{BU}	branched	N.A.	weekday	1 min	10
2	Model_{BU}	looped	N.A.	weekend	1 min	10
3	Model_{BU}	branched	N.A.	weekday	15 min	10
4	Model_{BU}	looped	N.A.	weekend	15 min	10
5	Model_{BU}	branched	N.A.	weekday	1 h	10
6	Model_{BU}	looped	N.A.	weekend	1 h	10
7	Model_{TD}	branched	DMP_{REF}	weekday	15 min	1
8	Model_{TD}	looped	DMP_{REF}	weekend	15 min	1
9	Model_{TD}	branched	DMP_{PS}	weekday	15 min	1
10	Model_{TD}	looped	DMP_{PS}	weekend	15 min	1
11	Model_{TD}	branched	DMP_{SIM}	weekday	15 min	1
12	Model_{TD}	looped	DMP_{SIM}	weekend	15 min	1

For each scenario, the Model_{TD} was run once and the system flow and water age at three locations were determined. The Model_{BU} was run 10 times with 10 different sets of stochastic water demand patterns. The resulting system flow (and corresponding DMP_{SIM}) is the averaged pattern of the 10 resulting patterns; the resulting water age at the three locations is determined by the average and the 95% confidence interval of the 10 simulations (mean $\pm$ twice the standard deviation). This 95% confidence interval is due to variation, not to uncertainty.

It was tested if 10 runs is enough to get a good view of mean and standard deviation of the residence times over the day. The difference between $(\mu + 2\sigma)$ of the residence time after N-1 simulations and $(\mu + 2\sigma)$ after N simulations reveals how large the effect of an extra simulation run is. To calculate the effect for $N = 10$, $(\mu + 2\sigma)_{10}$ is compared to $(\mu + 2\sigma)_9$. Because the 10 data points (i.e. calculated residence times) are the results of a Monte Carlo simulation, the order of the 10 data points is random. To account for this effect, the ten different subsets consisting of 9 data points are considered as possible results for $N = 9$. At each time step, the maximum difference (MD) between $(\mu + 2\sigma)_{10}$ and $(\mu + 2\sigma)_9$ as a percentage of μ_{10} is calculated:

$$MD = \frac{MAX(\mu + 2\sigma)_9 - (\mu + 2\sigma)_{10}}{\mu_{10}} \tag{6.1}$$

An MD of less than 5% is considered to be small enough to assume 10 simulation runs suffice.

Additionally, a Kolmogorov-Smirnov (KS) test was performed on the resulting residence time (10 data points) per time step (96 time steps at the hydraulic time scale of 15 min) on each of the three locations to verify that the data are normally distributed; the mean and variance of the normal distribution were estimated from the data. The null hypothesis was that the data are normally distributed and the test was performed at the 5% significance level. If the data are normally distributed, the mean and standard deviation can be used to demonstrate the results of several simulation runs.

Due to the fact that the instantaneous flow was logged per minute and not the average flow (Sec. 6.2.3), the total flow is not suitable for the model validation. Therefore, the DMPs are used for model validation. The modelled DMP_{SIM} and the measured DMP_{PS} and DMP_{REF} were compared. The diurnal pattern allows a visual assesment of how well the models resemble reality. To quantify the resemblance, the auto- and cross-correlation of the DMPs were considered. The cross-correlation between the DMPs shows how well the modelled DMPs fit the measured DMP_{REF}; cross-correlation can be established for different time lags, which shows if the modelled DMPs exhibit a delay with respect to the measured DMP_{REF}. The auto-correlation of the DMPs shows how variable the DMPs are.

The measured residence time at three locations and different times of day was compared to the modelled residence time (water age) in the two network modes. The difference between (the average of) the model and the measurement is expressed by the Mean Error (ME), Root Mean Square Error (RMSE), and coefficient of determination R^2. The absolute values of ME and RMSE are expressed in hours; the relative values are percentages of the measured residence times. Also, the percentage of the model values that differs less than 10 minutes from the measured value is calculated. For the $Model_{TD}$, this percentage is calculated for the average modelled values. For the $Model_{BU}$, this percentage is calculated for the average modelled values and for the 95% confidence interval of the 10 different runs.

6.3 RESULTS AND DISCUSSION – BENTHUIZEN AREA

6.3.1 Demand multiplier pattern

Figure 6.3 shows the diurnal patterns. On weekdays, DMP_{PS} does not show a very distinct morning peak while DMP_{REF} and DMP_{SIM} do. The start of low night use for DMP_{SIM} is later than for DMP_{REF} and DMP_{PS}. For weekend days, all DMP are similar. DMP_{REF} is more spiky, because it is only based on 2 to 3 days of measurements of instantaneous flows.

Figure 6.4 shows the auto- and cross-correlation for weekdays and weekend days. For weekdays, the auto-correlation of DMP_{REF} is represented slightly better by the auto-correlation of DMP_{SIM} than by the auto-correlation of DMP_{PS}. For example, at a time lag of 60 minutes, the auto-correlation of DMP_{REF} is 0.67, for DMP_{PS} it is 0.89 and for DMP_{SIM} it is 0.70 (Figure 6.4a). This means that the variability of the flow into the network is predicted better by DMP_{SIM} than by DMP_{PS}. For weekend days, DMP_{SIM} seems to have much better agreement to DMP_{REF} than DMP_{PS} does (Figure 6.4c).

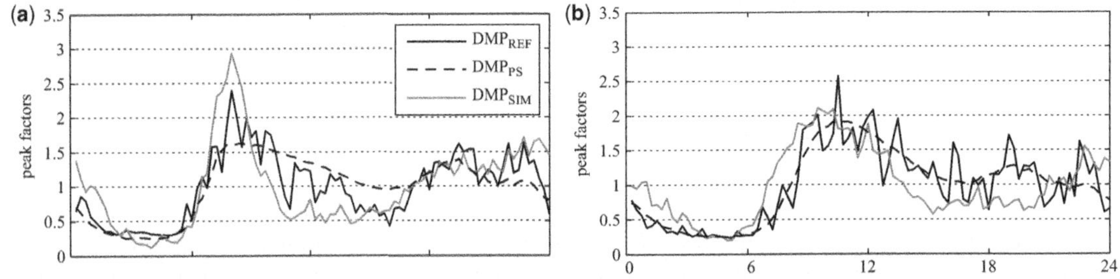

Figure 6.3 Measured and simulated normalised DMP at 15 minutes time step on a) weekdays and b) weekend days

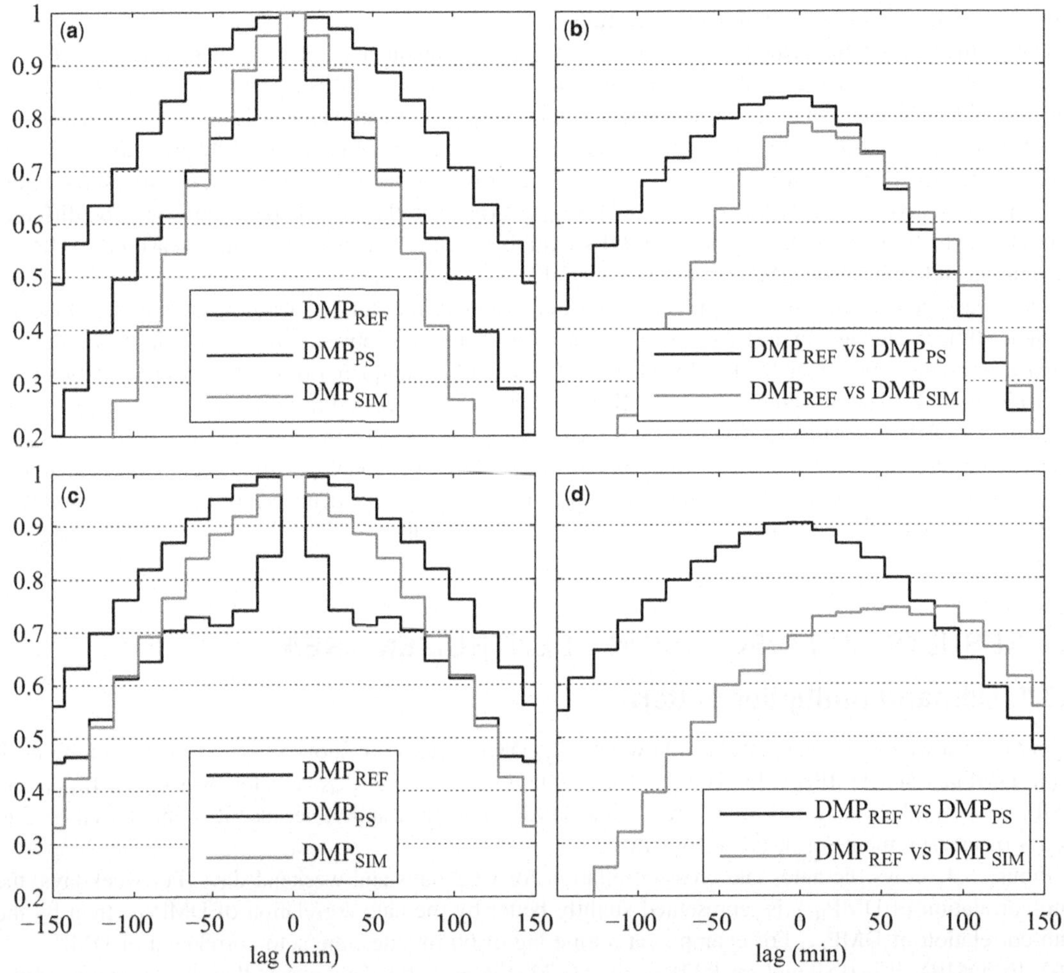

Figure 6.4 Auto-correlation (a, c) and cross-correlation (b, d) for different time lags, with time step of 15 minutes. Upper row (a, b) is for weekday patterns, lower row (c, d) is for weekend day patterns

The cross-correlation between DMP_{PS} and DMP_{REF} is slightly higher than the cross-correlation between DMP_{SIM} and DMP_{REF}. There is a maximum cross-correlation of 84% and 79% respectively on weekdays (Figure 6.4b) and 90% and 75% respectively on weekend days (Figure 6.4d). The maximum cross-correlation between DMP_{PS} and DMP_{REF} occurs at time lag = 0, thus DMP_{PS} has no delay with respect to DMP_{REF}. The maximum cross-correlation between DMP_{SIM} and DMP_{REF} occurs at time lag = 0 for weekdays and time lag = 90 minutes at weekend days. Thus, on weekends DMP_{SIM} runs 90 minutes ahead of DMP_{REF}. Figure 6.3 confirms that on the weekend the morning peak for DMP_{SIM} occurs too early.

6.3.2 Residence time – sensitivity analysis

At each time step, the maximum difference (*MD*, Eq. 6.-1) between $(\mu + 2\sigma)_{10}$ and $(\mu + 2\sigma)_9$ as a percentage of μ_{10} is calculated. Table 6.4 shows at how many time steps *MD* is smaller than 5%. It also shows the average of *MD* over all time steps. It shows that in the branched layout, 10 simulations runs lead to a stable result, i.e. a less than 5% difference of $\mu + 2\sigma$ of the residence time between the 9th and 10th simulation run. This conclusion can not be drawn for the looped layout at location 4. Especially with a short time step of 1 minute, 10 simulation runs does not give a stable result yet and more simulation runs are required.

Table 6.4 Relative maximum difference (*MD*) between $\mu + 2\sigma$ of 9th and 10th simulation run per time step

Location	Layout	Time steps where *MD* <5%			Average *MD*		
		1 min	15 min	1 h	1 min	15 min	1 h
2	Branched	100%	100%	100%	1.9%	1.7%	1.0%
3	Branched	100%	100%	100%	1.4%	1.3%	0.9%
4	Branched	100%	100%	100%	1.2%	1.2%	0.9%
2	Looped	100%	100%	100%	1.9%	1.7%	1.2%
3	Looped	99%	100%	96%	2.0%	1.9%	1.5%
4	Looped	83%	91%	96%	3.9%	3.6%	2.6%

A Kolmogorov–Smirnov (KS) test was performed on the resulting water age per time step on each location. The null hypothesis was that the data are normally distributed and the test was performed at the 5% significance level. At location 2 in the branched layout with a 15 min time step, the KS test showed that at 91 time steps (95%, Table 6.5) the null hypothesis could not be rejected. Therefore, it is assumed that in this case the resulting water age at each time step is normally distributed and that 10 runs are enough to get information on the mean and standard deviation. This assumption appears to be valid in all cases, except for location 4 in the looped layout, where the normal distribution can be confirmed for less than 90% of the time. In this case 10 runs of the $Model_{BU}$ may not be enough. For this case study, the number of simulation runs was limited to ten.

Figure 6.5 to Figure 6.7 show the modelled and measured residence time over the day for the different scenarios; Table 6.6 summarises the statistics. Depending on the network layout and the measurement location, the maximum residence time is reached between 5:00 and 9:00 a.m., which is related to the low night use. The fast decrease in residence time after the maximum is related to the peak in demand in the morning. The 95% confidence interval of the water age in the $Model_{BU}$ is the largest for location 4 in the looped network layout; this is due to local conditions and flow direction reversals (Figure 6.7f).

Table 6.5 Percentage of modelled water age results per time step that is normally distributed according to Kolmogorov-Smirnov test on 10 data points per time step

Location	Layout	Time scale		
		1 min	15 min	1 h
2	Branched	95%	95%	100%
3	Branched	95%	95%	100%
4	Branched	96%	98%	88%
2	Looped	95%	92%	96%
3	Looped	92%	93%	96%
4	Looped	82%	76%	50%

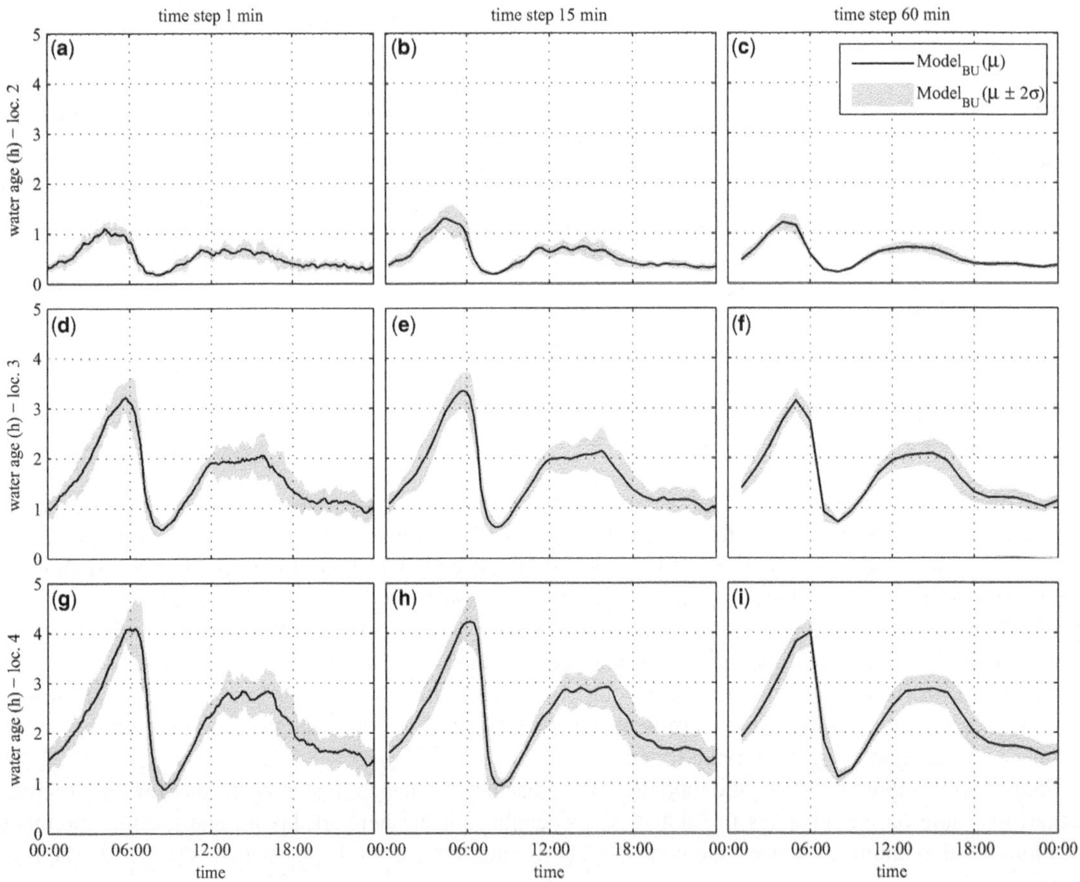

Figure 6.5 Modelled water age with Model$_{BU}$ in branched network layout (scenarios 1, 3 and 5) on locations 2 (a–c), 3 (d–f) and 4 (g–i) with a time scale of 1 min (a, d, g), 15 min (b, e, h) and 1 h (c, f, i)

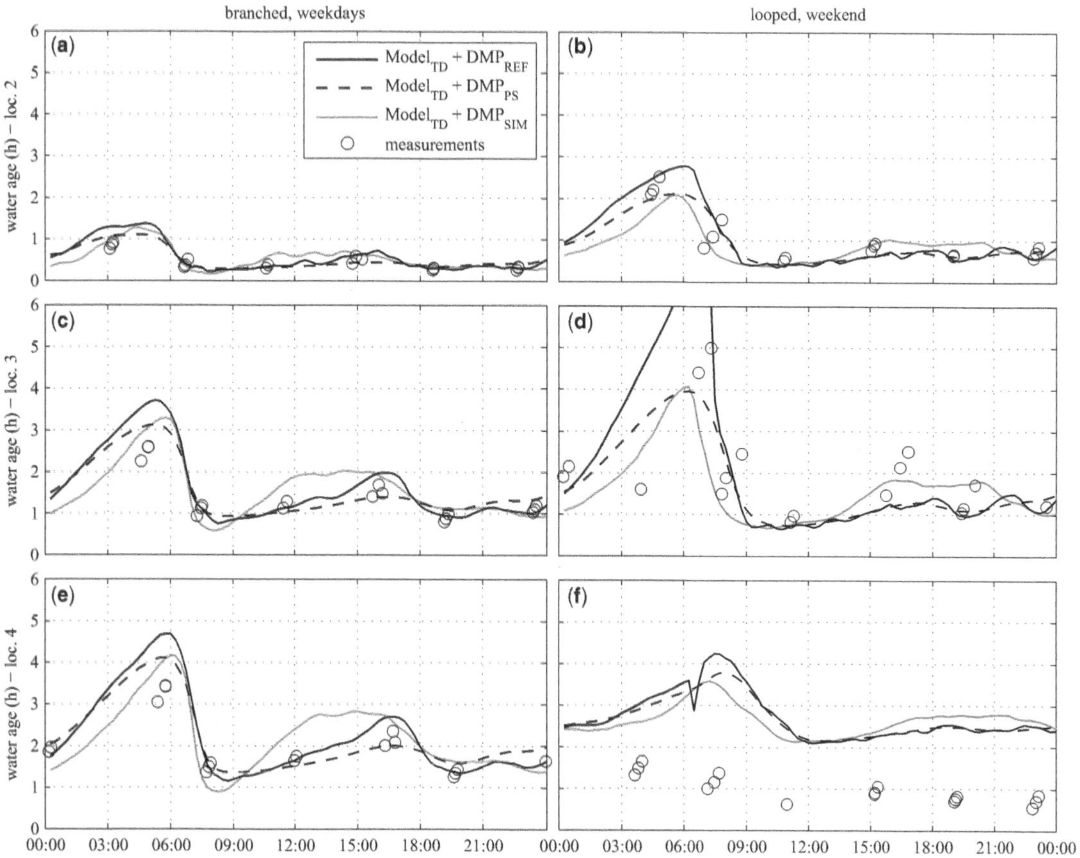

Figure 6.6 Measured residence time and modelled water age with Model$_{TD}$ and different DMP in branched (a, c, e; scenarios 7, 9 and 11) and looped (b, d, f; scenarios 8, 10 and 12) network layout on locations 2 (a–b), 3 (c–d) and 4 (e–f) with a time scale of 15 min

The effect of the model's temporal scale was determined by comparing the resulting water age from the Model$_{BU}$ with hydraulic time steps of 1 min, 15 min and 1 h (Figure 6.5). The difference in results is due to time averaging only. For determining residence time in this particular case, a 15-minute time scale is accurate enough. A shorter time step (1 minute) does not lead to a different 95% confidence interval of residence times. A longer time step (1 h) leads to too much time averaging; the minimum residence time is higher and the maximum residence time is lower than with a time step of 15 minutes. Therefore, this time scale was used in the remaining analysis of Figure 6.6 and Figure 6.7.

The effect of the model's spatial scale can be determined by comparing the resulting average water age from the Model$_{BU}$ with a hydraulic time steps of 15 min (Figure 6.5) and the resulting water age from the Model$_{TD}$ + DMP$_{SIM}$ (Figure 6.6). These two models vary in spatial correlation of the demands and in the location of the demand nodes, i.e. the Model$_{BU}$ has its demand nodes distributed along the pipes; the Model$_{TD}$ has its demand nodes at the ends of the pipes. The Model$_{BU}$ thus truly experiences demands per house and the ModelTD has its demands per consolidated node. In the branched layout (compare average of Figure 6.5b, e, h and black solid line in Figure 6.6a, c, e) there is no difference between the average of the $Model_{BU}$ and the $Model_{TD}$. In the looped layout (compare average of $Model_{BU}$ in Figure 6.7b, d, f and black solid line in Figure 6.6b, d, f) there is a small difference at locations 3 and 4: the $Model_{TD}$ results in a smoother line and shows a lower water age in the morning at location 3 and a higher water age in the morning at location 4. Spatial scale therefore has a limited effect on the mean and 95% confidence interval of calculated residence times at the measurement locations. The effect of spatial scale is eminent at the periphery of the network, where only a few homes are connected.

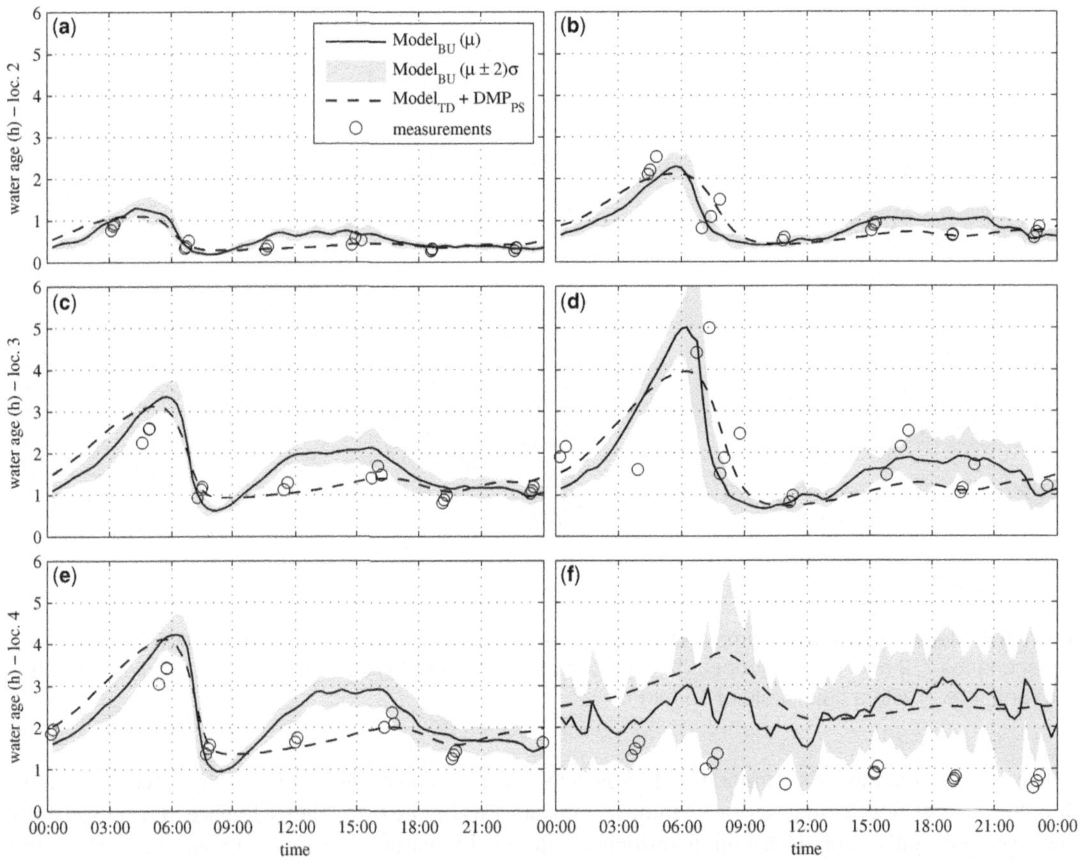

Figure 6.7 Measured residence time and modelled water age ($Model_{TD}$ + DMP_{PS} and $Model_{BU}$) in branched (a, c, e; scenarios 3 and 9) and looped (b, d, f; scenarios 4 and 10) network layout on locations 2 (a–b), 3 (c–d) and 4 (e–f) with a time scale of 15 min

The effect of the $Model_{TD}$'s DMP was determined by comparing the results from $Model_{TD}$ + DMP_{REF}, DMP_{PS} and DMP_{SIM} respectively (Figure 6.6). An effect of the different water demand patterns on water age was expected from Figure 6.3. The effect of the DMP is apparent as the three different DMPs lead to different water ages. The measured residence time is most often predicted best by the $Model_{TD}$ + DMP_{PS}, i.e. for location 3 and 4 in the branched layout and for location 2 and 3 in the looped layout (Table 6.6). Sometimes the $Model_{TD}$ + DMP_{SIM} works best, i.e. for location 2 in the branched layout

Table 6.6 Differences between measured residence time and modelled water age of Figure 6.6 and Figure 6.7

			$Model_{BU}$ (SIMDEUM)			$Model_{TD}$ (DMP_{PS})			$Model_{TD}$ (DMP_{REF})		
			L2	**L3**	**L4**	**L2**	**L3**	**L4**	**L2**	**L3**	**L4**
Branched	scenario			3			9			7	
	sample size		17	17	17	17	17	17	17	17	17
	ME	absolute (h)	0.05	0.22	0.25	0.05	0.15	0.21	0.10	0.26	0.33
		relative (%)	10.9	16.0	12.9	11.9	11.0	10.5	21.83	19.02	16.59
	RMSE	absolute (h)	0.11	0.40	0.53	0.13	0.33	0.42	0.21	0.53	0.65
		relative (%)	24.2	29.0	27.0	28.0	23.9	21.1	45.87	38.33	32.79
	R^2 (%)		72	50	42	63	66	64	N.A.	12	13
	Within 10 min deviation (%)	Compared to mean	88	18	6	88	35	41	76	59	59
		Compared to 95% c.i.	100	76	47	N.A.	N.A.	N.A.	N.A.	N.A.	N.A.
Looped	scenario			4			10			8	
	sample size		17	17	17	17	17	17	17	17	17
	ME	absolute (h)	−0.07	−0.37	1.46	−0.02	−0.28	1.76	0.04	0.07	1.89
		relative (%)	−7.0	−18.5	151.8	−2.2	−13.8	182.7	4.21	3.47	196.19
	RMSE	absolute (h)	0.37	1.01	1.64	0.35	0.86	1.86	0.41	1.19	2.03
		relative (%)	34.9	50.4	170.2	33.1	43.0	193.2	39.14	59.57	210.88
	R^2 (%)		66	22	N.A.	69	44	N.A.	57	N.A.	N.A.
	Within 10 min deviation (%)	Compared to mean	29	23	0	59	17	0	47	18	0
		Compared to 95% c.i.	88	47	35	N.A.	N.A.	N.A.	N.A.	N.A.	N.A.

and for location 4 in the looped layout (Table 6.6). The DMP that was actually measured does not lead to the best results when applied in the $Model_{TD} + DMP_{REF}$.

6.3.3 Residence time – model validation

The difference between the conventional approach ($Model_{TD} + DMP_{PS}$, because DMP_{REF} would not usually be available to the drinking water company Dunea) and the new approach ($Model_{BU}$) of demand modelling is shown in Figure 6.7 and Table 6.6. The two models predict the residence time with comparable ME and RMSE and R^2. Both models predict the residence time in the branched layout with an ME and RMSE of less than 30%. Both models perform poorly for the looped layout (RMSE > 30%) and especially for location 4 (Figure 6.7f) where they significantly overestimate the water age. The values of ME, RMSE and R^2 only have a meaning in comparing the two models. The absolute values are not easy to interpret because the measured residence times at specific moments on the day are different for consecutive days and therefore show variance. However, the average of the model is compared to the variable measured data. The 95% confidence interval of the $Model_{BU}$ presents many more data points within 10 min from the measured residence time than the average of both the $Model_{BU}$ and the $Model_{TD}$. This shows the added value of the $Model_{BU}$.

The extra demand of 560 L/h (3*40 L/h + 400 L/h) of the monitor systems is relatively large compared to the average residential demand of 1884 L/h (314 L/home.day * 144 homes / 24 h). At location 4, the odds are even more extreme. The influence of the constant demand is, therefore, large. However, since the residential demand is very variable, the influence of the residential demand in the measurements as well as in the model is still substantial. The application of SIMDEUM can be validated with this model. In the next case study, SIMDEUM will be validated in a network with much smaller continuous demands.

6.4 INTERMEDIATE CONCLUSIONS FROM BENTHUIZEN STUDY

The Benthuizen measurements showed how to do a tracer test and how to estimate the required duration of the NaCl dosage. The Benthuizen sensitivity test showed that for the stochastic bottom-up approach 10 simulation runs were sufficient to determine the average and 95% confidence interval. It also demonstrated that the required detail in spatial scale was limited demands in the model can be concentrated on pipe ends, i.e. there is no need to put the demand nodes in the model exactly at the connection point of the homes. Furthermore, it showed that a 1-hour time step was too long and a detail of a 1-min time step was not required. A 5- to 15-minute time step should suffice.

The Benthuizen model validation provided an extra validation of SIMDEUM with a demand pattern of 140 homes. It also showed that the bottom-up approach predicts the residence time at least as well as the top-down approach. Furthermore, it demonstrated that the stochastic approach gives insight into the variation of demand and the variation in residence times.

6.5 METHODS AND MATERIALS – ZANDVOORT AREA

A distribution network of about 10 km of mains and 1000 homes was selected as a second test area. In this network, the total flow was measured and a tracer study was performed to determine the residence time at four locations in the network. An "all pipes" hydraulic model was constructed with two methods of demand allocation: one with a top-down approach of demand allocation with one unique DMP, and another with a bottom-up approach of demand allocation of individual stochastic demand patterns. The model results were compared to the measured flow and residence time.

6.5.1 The network

The selected network is situated in the Dutch town of Zandvoort, along the sea. The network was built in the 1950s-1960s and consists of 3.5 km of PVC pipes, and 5.7 km of lined cast iron pipes (Figure 6.8, Table 6.7). It supplies about 1000 homes, 2 hotels and 30 beach clubs. The area is supplied from one point with a fixed head through a booster pump; there are no tanks in the network.

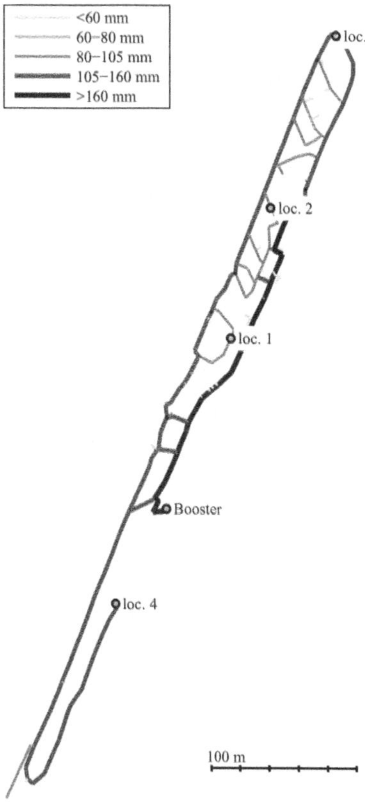

Figure 6.8 Network layout

Table 6.7 Pipe diameters and materials in the network

Diameter (mm)	Length (km)		Volume (m^3)
	CI	PVC	
<100		1.4	7.8
100	1.3	0.6	14.8
150	3.4	1.1	79.1
180		0.4	12.9
225	0.9		35.0
total	5.6	3.5	149.7

The water use in the network was determined from the historic flow patterns at the booster station, as measured by the Provincial Water Company Noord-Holland (PWN) and is, on average, 24 m^3/h. The domestic water demand is 70% of the total demand; the leakage in this network is not known. As leakage in the Netherlands is generally very low (2–4%), no leakage is assumed in this network. The average residence time is 149.7 m^3 / 24.0 m^3/h = 6.2 h (Table 6.7, Table 6.8).

The drinking water is distributed without any disinfectant, as is common in the Netherlands. A tracer study with NaCl was done between 2 September and 20 October 2008.

6.5.2 Measurement setup for the tracer study

The tracer was added at the booster station. In the network, four measurement locations were selected (Figure 6.8). Location 1 and location 2 are located near apartment buildings on a Ø100 mm PVC and Ø100 mm CI pipe. Location 3 is situated in the basement of a hotel. Location 4 is situated in the basement of a small apartment building of 15 residences.

As in the Benthuizen tracer test (Section 6.2), sodium chloride (NaCl) was used as a tracer and the electrical conductivity was measured. From these measurements, the residence time was determined. At the booster location, NaCl was dosed to a fixed concentration in order to raise the electrical conductivity (EC, in mS/m) by a measurable amount: EC ≈ 57 mS/m without dosage, and EC ≈ 68 mS/m with dosage. The tracer was dosed in pulses of 3 hours on and 20 hours off. This means that, per day, one positive and one negative step input were induced. A three-hour pulse should be long enough to have a distinctive start and end time at the measurement locations, given the expected longitudinal spreading. A 20-hour window of not dosing the tracer ensured that enough NaCl was available for a 7-week test. The tracer dosing schedule thus shifted the start time of the pulse by 1 hour every day. Hence, during the seven- week field study, each hour of the day was used twice as a starting time for the tracer test.

In order to reach a fixed concentration, the incoming flow was measured (Tokimec UFP-10) and the dosage was controlled (Figure 6.9). A solute of 220 g/l NaCl was added with a maximum pump capacity of 20 l/h during the maximum demand of 60 m^3/h, and an average dosage capacity of 8 l/h during the average demand of 24 m^3/h. This led to a rise of 11 mS/m. Dosings of 3 hours per 23 hours over 50 days means that ca. 275 kg of NaCl was used in the tracer test. The booster pumps ensured a fixed head and a constant concentration of the tracer in the water over the pipe, i.e. good mixing.

The (average) flow was logged every minute for 16 full days of flow measurements at the booster station and 11 full days of flow measurements at location 3. The measured flows were time averaged over 5 minutes and converted into dimensionless average demand multipliers, which are denoted $DMP_{booster}$ and DMP_{hotel}. The flow measurements at location 3 showed that the daily demand varied between 2.21 and 4.32 m^3/h and can be described by a Weibull distribution. Based on 11 days, the parameters a and b of the Weibull distribution have been estimated at $a = 3.247 \pm 0.4$ and $b = 4.741 \pm 1.7$.

At all four locations, the EC was measured (LIQUISYS M CLM223). The measurements required a continuous 40 l/h extraction. The EC measurement at the booster station was not logged, instead the dosage regimen was recorded.

The residence time between the booster station and locations 1, 2, 3 and 4 was determined using the time between the centres of the ascending and descending tails of the EC pulses at around 61 mS/m, which are denoted as τ_a and τ_d, respectively, and by the time between the centroids of the pulses, denoted by τ_c. The centroid was determined by the weighted mean between τ_a and τ_d. Figure 6.10 shows that, at around 7:00 o'clock the residence time between the booster and location 1 was equal to 3.3 h (ascending tail); at ca. 7:40, the residence time between the booster and location 1 was equal to 2.6 h (centroid) and at about 8:30 the residence time between the booster and location 1 was equal to 1.8 h (descending tail). The residence

Table 6.8 Demands in $Model_{TD}$ and $Model_{BU}$, # units with given base demand are reported; these are not all connected to an individual demand node

Demand category	**$Model_{TD}$**				**$Model_{BU}$**			
	#	**Q_{base} (m^3/h)**	**Q_{total} (m^3/h)**	**Pattern**	**#**	**Q_{base} (m^3/h)**	**Q_{total} (m^3/h)**	**Pattern**
small beach club	21	0.05	1.05	$DMP_{booster}$	21	0.05	1.05	$DMP_{booster}$
large beach club	11	0.10	1.10	$DMP_{booster}$	11	0.10	1.10	$DMP_{booster}$
residence type A	100	0.015	1.50	$DMP_{booster}$	869	N.A.	10.20 – 10.60	SIMDEUM res. type A
residence type B	210	0.02	4.20	$DMP_{booster}$	210	N.A.	3.45 – 3.75	SIMDEUM res. type B
apartment building	25	$\geq$*0.30*	*10.30*	$DMP_{booster}$		N.A. (*moved to residence type* A)		
apartment building, loc. 4	1	0.20	0.20	$DMP_{booster}$				
NH Hotel, loc. 3	1	3.247	3.247	$DMP_{booster}$	1	Weibull distributed ($a = 3.247$, $b = 4.741$)	1.73 – 4.20	DMP_{hotel}
Beach Hotel	1	1.783	1.783	$DMP_{booster}$	1	1.783	1.783	DMP_{hotel}
Palace Hotel	1	0.50	0.50	$DMP_{booster}$	1	0.50	0.50	DMP_{hotel}
Measurement location	4	0.04	0.16	Constant demand	4	0.04	0.16	Constant demand
Total			24.04				20.29 – 22.40	

time varies over the day and between days. This variation is considered in both the measurements and the hydraulic model.

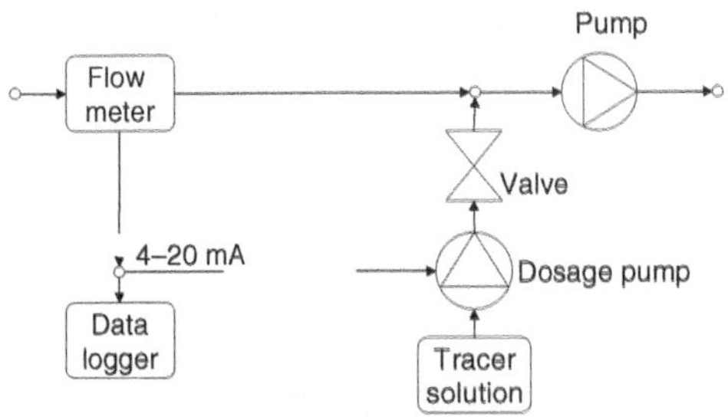

Figure 6.9 Measurement setup for adding tracer solution

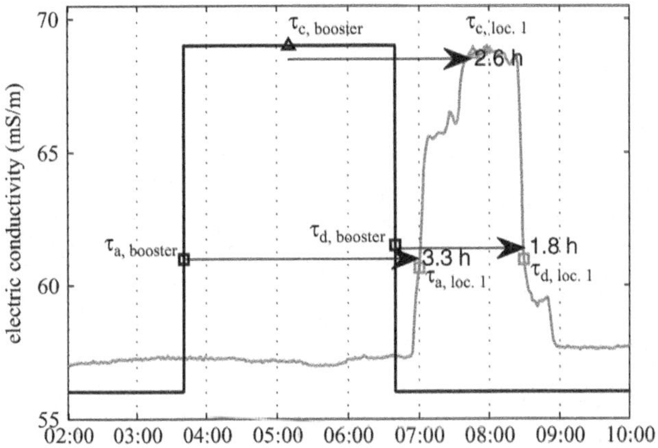

Figure 6.10 Reconstructed EC at booster locations and measured EC at location 1(Wednesday, 3 September 2008) plus associated residence times

6.5.3 Hydraulic model and demand allocation

EPANET 2.0 (Rossman 2000) was used as a hydraulic network model solver. The wall roughness of the Cast Iron pipes was set to values between 0.2 and 1 mm, as they were provided by the water company. Basically, two models were constructed that are distinguished by demand allocation. Model_{TD} is the model with the top-down approach of demand allocation; Model_{BU} is the model with the bottom-up approach of demand allocation.

The measurement locations 1, 2, 3 and 4 were assigned a continuous extraction of 40 l/h. No pressure dependent demands or leaks were introduced in the model. The water demand allocation was conducted as follows:

1. In Model_{TD} an identical DMP ($\text{DMP}_{\text{booster}}$, Figure 6.11) was allocated to all demand nodes with a correction factor to account for the average demand. This correction factor is the base demand

and it was assigned according to the demand category (Table 6.8). Measurement location 4 is at the end of a branch and EC measurements were done at that apartment building. Therefore, the apartment building at measurement location 4 was assigned a better estimate of base demand. The base demand was calculated for 15 homes of residence type A (Table 6.9; 15 homes × 2.3 persons/home × 129.3 L/person.day ≈ 0.2 m^3/h). A demand node may serve multiple homes or beach clubs.

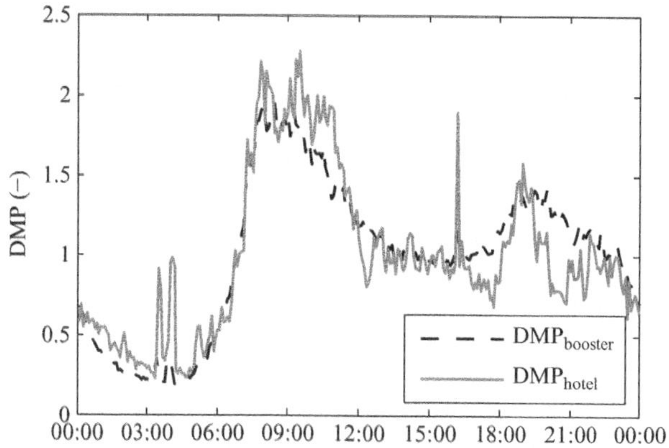

Figure 6.11 Demand multiplier patterns as used in the $Model_{TD}$ and $Model_{BU}$, see Table 6.8. Measured average demand at booster is 23 m^3/h; measured average demand at NH Hotel is 3 m^3/h

Table 6.9 Specific input data into SIMDEUM; data of Zandvoort Boulevard (1040 homes) in 2003-2007 (CBS)

		Zandvoort Boulevard	SIMDEUM residence type A	SIMDEUM residence type B
Households	One person households	56%	34%	20%
	Households without children	34%	30%	34%
	Households with children	10%	36%	46%
	Average household size	1.6	2.3	2.7
Age distribution	0 to 12 years old	4.8%	15%	4.8%
	12 to 21 years old	3.7%	10%	3.7%
	21 to 65 years old	62.5%	63%	62.5%
	65 years and older	29%	12%	29%
Water using appliances	WC		No 6L cisterns	No 6L cisterns
	Outside tap		No	Yes, frequency of 0.7 /day
Average water use (L per person per day)			129.3	149.2

2. In $Model_{BU}$ different demand patterns were assigned to different demand category nodes (Table 6.8).
   - To each residential demand node (residence types A and B), a unique stochastic water demand pattern was assigned. The stochastic water demand patterns were obtained from the end-use model SIMDEUM (Chapter 3), see Sec. 6.5.4. The residential water use accounts for 68% of total water use.
   - SIMDEUM currently provides only residential demand patterns. SIMDEUM was applied to predict water demands of offices, hotels and nursing homes (Chapter 4). These specific applications have shown that SIMDEUM can be extended to non-residential water demand. However, not all model parameters for the non-residential water demand have been identified. Since some flow measurements were collected, these were used instead of SIMDEUM demand patterns.
   - Regarding the hotel demand nodes, the measured DMP_{hotel} (Figure 6.11) and base demand as in the $Model_{TD}$ were assigned for two hotels; for the hotel at location 3, a base demand according to the Weibull distribution was used. Two hotels thus have a deterministic demand pattern; they account for 9.5% of the total water use. The largest hotel, which acocunts for 13.5% of the total demand, has a deterministic demand pattern but a stochastically determined base demand.
   - For the beach club demand nodes, the same $DMP_{booster}$ and base demand as in the $Model_{TD}$ were assigned. The beach clubs have a deterministic demand pattern; they account for 9.0% of the total water demand.

The hydraulic and pattern time step was set to 15 minutes in the $Model_{TD}$ and to 5 minutes in the $Model_{BU}$; the EPANET water quality time step was set to 1 minute in both models.

6.5.4 Water demand pattern generation

The end-use model SIMDEUM (Chapter 3) was used as a water demand pattern generator. SIMDEUM input consists of information on the number and age of residents, the residents' time use and possession of and behaviour with respect to water-using appliances. Generic Dutch data were used for the water-using appliances (Kanne 2005a) and time use (SCP 1995). Specific data about Zandvoort were used for household composition and water-using appliances (Table 6.9). The type A residences (often apartments, mainly in the north) do not have a garden and no outdoor water use. In the south, type B residences (villas) are found. The census data of Zandvoort were not used because in the measurement period (late summer) it was expected that people who were not inhabitants would also be occupying the homes because this part of Zandvoort atracts a lot of tourists. Also, the residence type B is larger than average and can, therefore, be occupied by more people. The residence type B has a garden and it is assumed that they have outdoor water use because in the summer the gardens are likely to be watered. To summarise, only the data of Table 6.9 were specific, all other input data for SIMDEUM were found from the average Dutch data (Chapter 3).

For each of the 869 residence type A and 210 residence type B homes, 14 unique water demand patterns on a time scale of 1 s were generated; 10 patterns for weekdays and 4 patterns for weekend days. The patterns were then temporally aggregated (over 5 min) and spatially aggregated (to the number of connections per demand node). This led to 14 DMPs for each demand node. The Benthuizen study showed that 10 different sets of demand patterns are enough to assume a normal distribution and to allow the calculation of the mean and 95% confidence interval (Section 6.3.2).

Next, 14 copies of the EPANET.inp file were created and then adjusted. To the list of patterns, demand pattern IDs were added plus 288 demand multipliers (pattern time step was 5 min) that were generated by

SIMDEUM, the demand pattern from the hotel (DMP_{hotel}) and from the booster ($DMP_{booster}$). The average of the SIMDEUM demand multipliers is not equal to 1, but to the actual average residential demand. Instead of dividing the flows by their average to get a DMP with an average of 1 and assigning the average flow to the demand nodes, the DMPs can be viewed as representing the flow in L/s (with an average unequal to 1). Each junction that represents residential demand was thus assigned a base demand of 1 L/s and a unique demand pattern ID. The junctions that represent a hotel were given the base demand from the water meter readings (Table 6.8). The hotel at measurement location 3 was assigned a variable base demand randomly drawn from the Weibull distribution (Table 6.8). The junctions that represent a beach club were given the base demand from Table 6.8.

6.5.5 Model validation

The $Model_{TD}$ was run once and the system flow and water age at four locations were determined. The $Model_{BU}$ was run 14 times with 14 different sets of stochastic water demand patterns (10 weekdays, 4 weekend days) and 14 different base demands at the NH hotel (location 3). The resulting system flow (Q_{SIM}) is the averaged pattern from the 14 resulting patterns; the resulting water age at the three locations was determined by the average and the 95% confidence interval of the 14 simulations. This 95% confidence interval is due to variation, not to uncertainty.

The resulting system flow Q_{SIM} was compared with the average measured flows $Q_{booster}$ on a time scale of 5 minutes. The measured residence time at four locations and different times of day was compared to the modelled water age in the network. The difference between model and measurement is expressed by the Mean Error (ME), Root Mean Square Error (RMSE), and coefficient of determination R^2. The absolute values of ME and RMSE are expressed in hours; the relative values are percentages of the measured residence times. Also, the percentage of the model values that differ less than 10 minutes from the measured value was calculated. For the $Model_{TD}$, this percentage was calculated for the average modelled values. For the $Model_{BU}$, this percentage was calculated for the average modelled values and for the total of the 14 different runs.

6.6 RESULTS AND DISCUSSION – ZANDVOORT AREA

6.6.1 Diurnal flow pattern

Figure 6.12 shows the measured and modelled diurnal pattern. The two patterns show good resemblance. For the beach clubs (9% of total water demand), the booster pattern was used (Figure 6.11). For the hotels (23% of total water demand), the hotel pattern was used (Figure 6.11), which is similar to the booster pattern. The remaining 68% of the total water demand originates from the residential SIMDEUM patterns. The modelled flow pattern has a more distinct morning peak than the measured pattern. The simulated pattern shows a later decline to low night use than the measured flow. The small peaks around 4:00 a.m. are due only to the hotel's demands. The modelled flow shows these peaks because the beach clubs have them in the applied $DMP_{booster}$ and the hotels have them in the applied DMP_{hotel}. These two peaks are most likely related to cleaning. The modelled and measured minimum flows are equal. This suggests that leakage is indeed very limited and can be ignored.

Figure 6.13 shows that the modelled flow pattern has a smaller auto-correlation than the measured flow pattern. The cross-correlation between $Q_{booster}$ and Q_{SIM} is at a maximum of 0.9, at a time lag of 0, i.e. there is no delay. The morning peak of Q_{SIM} coincides with the morning peak of $Q_{booster}$ (Figure 6.12).

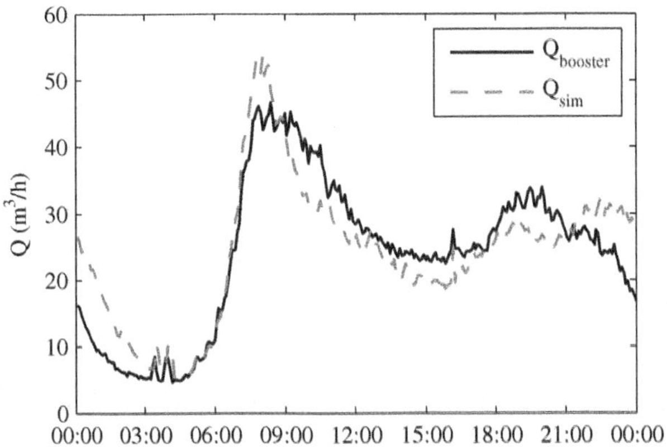

Figure 6.12 Measured ($Q_{booster}$) and simulated (Q_{SIM} from $Model_{BU}$) flows on a 5-minute time scale

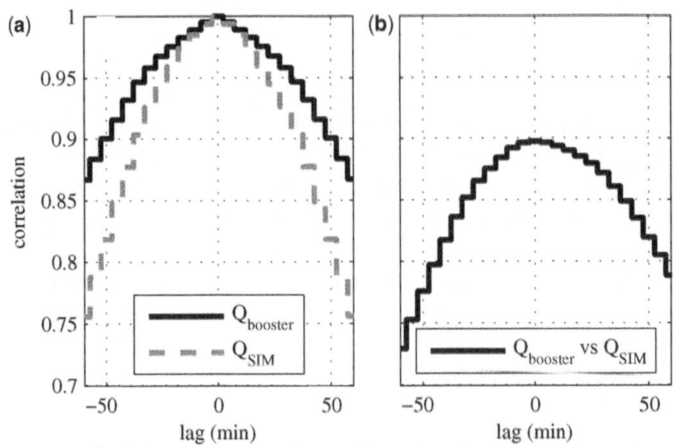

Figure 6.13 a) Auto-correlation of $Q_{booster}$ (measured) and Q_{SIM} (from $Model_{BU}$) and b) cross-correlation between $Q_{booster}$ and Q_{SIM} for different time lags on a 5-minute time scale

6.6.2 Residence time

At locations 1, 2 and 4 the measured EC resembles the rectangular pulse leaving the booster station, and the centres of the ascending and descending tails and the weighted means between those centres can easily be determined (Figure 6.14). Each pulse at the booster station led to 3 measured residence times at those locations: 138 measurement points, in total, at each location. At location 3, the pulse changed shape due to mixing and dispersion, and often more than one pulse can be seen (Figure 6.14). The residence time could only be determined at the ascending tail of the pulse. Each pulse at the booster station led to 1 measured residence time at location 3: 46 measurement points in total.

Figure 6.15 shows the measured residence time and modelled water age over the day at the four measurement locations; Table 6.10 summarises the statistics. Depending on the measurement location, the maximum residence time is reached around 7:00 a.m., which is related to low night use. The fast decrease in residence time after the maximum is related to the peak in demand in the morning. The 95% confidence interval of the water age in the $Model_{BU}$ is the largest for location 4 because there are only 15 homes present at that measurement location. The individual behaviour of the people in those homes has a large effect on demand and thus on residence time.

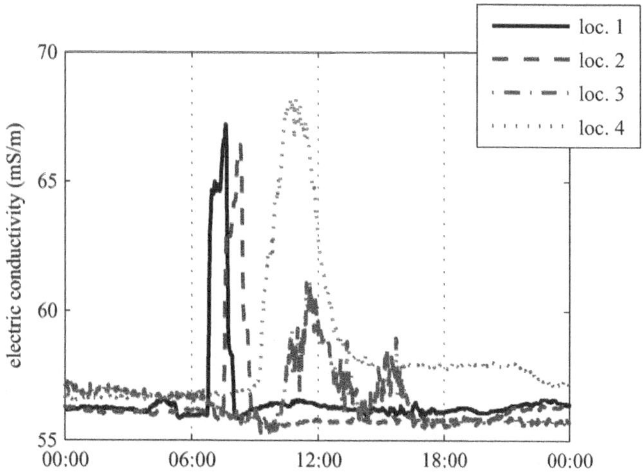

Figure 6.14 Measured EC at locations 1–4 at Thursday, 4 September 2008. The EC at location 4 has a 24-hour delay and is connected to the pulse that left the booster station on Wednesday, 3 September 2008

The average and 95% confidence interval of the water age from the $Model_{BU}$ and the water age from the $Model_{TD}$ with $DMP_{booster}$ were compared with the measured residence times. The two models predicted the residence time well, with an ME and RMSE of less than 30%. The $Model_{BU}$ shows lower ME and RMSE than the $Model_{TD}$. The 95% confidence interval of the $Model_{BU}$ presents much more data points within 10 min from the measured residence time than the average of both the $Model_{BU}$ and the $Model_{TD}$. The calculated R^2 is not a meaningful value for either model, and therefore it is not shown in the table.

6.6.3 Pulse shape

The centroids and ascending and descending tails of the pulses were converted into residence times, and thus the measurements and models were compared. Another possibility is to compare the pulse shapes. However, it is not easy to compare the shapes of all pulses. It may be required to first shift the pulses such that the beginning of the pulses (or the centres of the pulses) coincide. Next, it must be decided for what duration the pulses are to be compared, i.e. how long does the measured pulse endure? Then, the R^2 between measured and modelled pulse can be determined. In this study, only the pulse shape of one day at one measurement location was considered.

The pulse shape of the NaCl varied a lot between days, especially at measurement location 3. Water reaches location 3 via the pipes east of the connection pipe and via the west. Depending on the water use, which differs over the day and between days, the received NaCl may appear as more than one pulse. Since at each day the pulse leaves the booster station at a different time of the day, each measurement day shows different measured pulse shapes.

Figure 6.16 shows the measured pulse shape at location 3 on Wednesday, 14 and Thursday, 15 October 2008. The concentration is normalised with respect to the concentration at the booster station. The NaCl

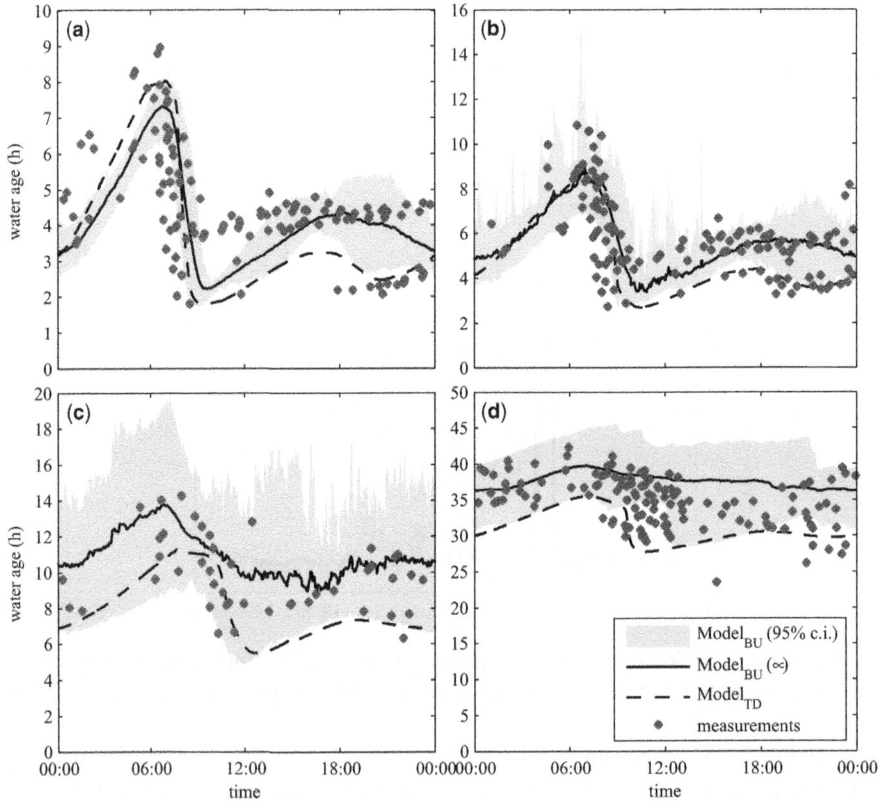

Figure 6.15 Measured residence time and modelled water age at location a) 1; b) 2; c) 3 and d) 4. N.B. the 95% confidence interval is due to variation, not to uncertainty

Table 6.10 Average absolute and relative differences between measured residence time and modelled water age

		Model$_{BU}$				Model$_{TD}$			
		loc. 1	**loc. 2**	**loc. 3**	**loc. 4**	**loc. 1**	**loc. 2**	**loc. 3**	**loc. 4**
Sample size		126	138	46	135	126	138	46	135
ME	absolute (h)	−0.14	−0.06	0.17	2.06	−0.27	−0.48	−1.45	−4.41
	relative (%)	−3.14	−0.99	1.76	5.91	−5.89	−8.23	−14.73	−12.63
RMSE	absolute (h)	1.42	1.42	1.65	4.05	1.85	1.77	2.47	5.68
	relative (%)	31.09	24.08	16.83	11.61	40.50	30.18	25.17	16.28
Within 10 min deviation (%)	Compared to mean	9.52	7.97	13.04	0.74	3.97	14.49	0	0.74
	Compared to 95% c.i.	64.29	79.71	100.0	77.78	N.A.	N.A.	N.A.	N.A.

pulse left the booster station on 14 October at 8:40 and arrived at the measurement location as two pulses. The figure also shows the pulse shape of the $Model_{TD}$ and of one (out of 10 weekdays) $Model_{BU}$. The pulse in the $Model_{TD}$ arrived hours earlier than the measured pulse and was not split. The pulse in the $Model_{BU}$ is split into two, and resembles the shape of the measured concentration quite well. The other 9 runs of the $Model_{BU}$ resulted in pulses much more like the one of the $Model_{TD}$, i.e. the resemblance with the measured data is not very good.

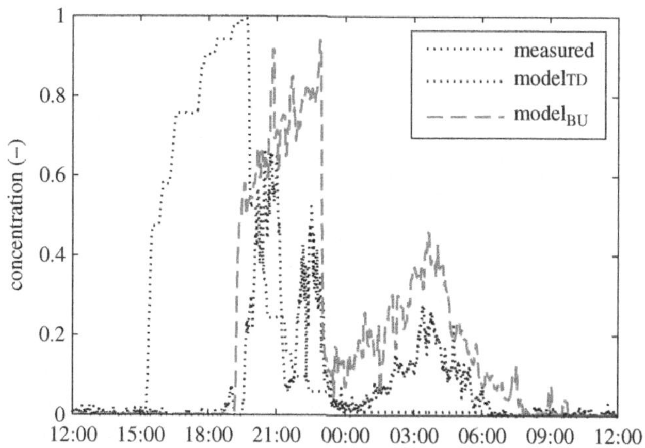

Figure 6.16 Measured and modelled pulse at measurement location 3 on 14-15 October 2008

6.7 INTERMEDIATE CONCLUSIONS FROM ZANDVOORT STUDY

In the Zandvoort tracer test, the lessons from the Benthuizen measurements were applied. In the Zandvoort model, the employed number of simulations and spatial and temporal scales were based on the sensitivity analysis of the Benthuizen model.

The Zandvoort model validation provided an extra validation of SIMDEUM with a demand pattern of 1000 homes. However, the total flow pattern also contained deterministic demand patterns of hotels and beach clubs. The model comparison also showed that the bottom-up approach predicts the residence time at least as well as the top-down approach. Furthermore, it demonstrated that the stochastic approach gives insight into the variation of demand and the variation in residence times.

The EC-pulse leaving the booster station on Thursday, 4 September (between 2:40 and 5:40) led to the EC-pulses of Figure 6.14 at locations 1, 2 and 3. The EC-pulse leaving the booster station on Wednesday, 3 September (between 3:40 and 6:40) led to the EC-pulses of Figure 6.14 at location 4. The maximum EC on Wednesday was higher than on Thursday. The maximum EC at location 4 was always slightly lower than at locations 1 and 2; the EC-pulse at location 4 was always much wider. This means that dispersion occurred.

6.8 GENERAL DISCUSSION

A tracer test was done in two areas. For these areas, an "all pipes" hydraulic model was constructed with two methods of demand allocation: one with a top-down approach of demand allocation with a common DMP and another with a bottom-up approach of demand allocation of individual and unique stochastic demand patterns. The Benthuizen area was a small area (144 homes) where extra flow was induced in order to limit the occurrence of laminar flow and thus limit dispersion; the Zandvoort area was larger (100 homes) and experienced normal dispersion. The Benthuizen case demonstrated how to do the measurements, how to compare the model with the measurements and what the best required spatial and temporal aggregation was. The Zandvoort case showed that the new approach of demand allocation also works in a larger network with dispersion. Figure 6.14 shows longitudinal spreading for both location 3 and location 4. As 30% of the pipes experienced laminar flow for more than 50% of the time, dispersion is an issue in this network. In the pipes leading to location 4, flows were laminar during most of the day; in these laminar flow pipes, the residence time was approximately 24 hours. In the part of the network around location 3, the flows were also mainly laminar. As the residence time was derived from the NaCl pulse, the measured residence time includes dispersion. Therefore, when the measured residence time is compared to model results, the model should also incorporate dispersion. It would be interesting to validate an Advection-Dispersion-Reaction model (Li et al. 2006; Tzatchkov et al. 2002) in this network.

The models were first compared on total flow. In the Benthuizen case, the $Model_{TD}$ (DMP_{PS}) had a cross-correlation of more than 80% with the measured flow (DMP_{REF}). The fact that the supply area is mainly residential contributes to this. The $Model_{BU}$ (DMP_{SIM}) had a cross-correlation of more than 75% with the measured flow (DMP_{REF}). The auto-correlation of the $Model_{BU}$ (DMP_{SIM}) resembles the auto-correlation of the measured flow (DMP_{REF}) better than the auto-correlation of the $Model_{TD}$ (DMP_{PS}) does. This suggests that, in terms of variability over the day, the $Model_{BU}$ behaves more like the actual flows than the $Model_{TD}$ does. In the Zandvoort case, the average measured flow pattern ($Q_{booster}$) was compared with the flow pattern from the simulated water demand patterns of the $Model_{BU}$ (Q_{SIM}); the flow pattern used in the $Model_{TD}$ was equal to the measured flow patterns. The cross-correlation between Q_{SIM} and $Q_{booster}$ was almost 90%.

Next, the models were compared with respect to residence time. The $Model_{BU}$ predicts the average residence time equally well (Benthuizen) or slightly better (Zandvoort) than the $Model_{TD}$. The $Model_{BU}$ provides information on the variability of the residence time, and thus the 95% confidence interval of the $Model_{BU}$ leads to better predictions of the residence time than the $Model_{TD}$ does. The model's sensitivity is mainly related to the variability of residence times, not so much to the average residence times.

The Benthuizen case showed that the difference between the results of the $Model_{BU}$ and the $Model_{TD}$ is mainly due to the used demand patterns and the stochastic nature of the bottom-up approach. The time scale and average DMP have a small effect on predicted residence time. The temporal scale in this particular case can be as long as 15 minutes without losing information; this may be due to the fact that dispersion could be neglected here. The model of a branched network layout is less sensitive to the type of demand allocation and less sensitive to different temporal and spatial scales than the model of a looped network layout is.

In the Zandvoort case, the hotels and beach clubs did not have a residential demand pattern assigned to them; instead, measured DMP were used. The beach clubs had an average demand of 9% of the total network demand; the average demand of the hotels is 23% of the network demand. With the extension of SIMDEUM to also simulate non-residential demands, the network model could be constructed without measured water demands.

The bottom-up modelling approach is probabilistic in nature and offers a new perspective for assessing water quality in the drinking water distribution system. The Benthuizen case showed that, especially at location 4 in the looped network layout, the variability of residence time between days is expected to be very high, with the maximum residence time 2.5 times as large as the average residence time, or the minimum residence time 2.5 times as small as the average residence time. This suggests that it would be very difficult to use the tracer measurements at this location for calibration purposes. The averaging of water demand and water age predictions may lead to misinterpretation of water quality data. Calibration of the top-down model of Zandvoort was attempted (Pardo Picazo 2009). However, calibration on pressure appeared to be impossible due to small head losses. Calibration of demands on the tracer measurements appeared to be difficult because the demand was so variable.

The stochastic approach of hydraulic modelling gives insight into the variability of residence times as an added feature beyond the conventional way of modelling. The conventional $Model_{TD}$ has a higher auto- and cross-correlation of flows than the actual flows in the network. This results in the $Model_{TD}$ underestimating the flow direction reversals, stagnant flows and, thus, maximum residence times. Machell et al. (2009) have argued that the maximum residence time is much more important than the average residence time. Therefore, the $Model_{BU}$ has benefits in determining residence time.

6.9 CONCLUSION

A bottom-up approach of demand allocation (i.e. water demand patterns are modelled per individual home and, subsequently, the individual water demand patterns are summed to obtain the water demand patterns at demand nodes) leads to a total flow that is predicted at least as well as the flow from the commonly used top-down approach model. The individual demand patterns are obtained from the end-use model SIMDEUM without the need for any flow measurements. Some specific census data were collected and used as input to SIMDEUM; most of the input data can be re-used from earlier studies.

A bottom-up approach leads to good results in predicting residence time in a small- and intermediate-scaled distribution network. There is no need for measuring water demand patterns, nor for calibrating demand based on water quality parameters.

The water demand patterns are constructed per individual home and on a per second basis. For the purpose of residence time prediction, it is acceptable to use time averaging, a hydraulic time step of 5 to 15 minutes, and 'spatial-averaging' by summing numerous individual water demand patterns into one demand node.

A stochastic approach in demand and water quality modelling results in more insight into the variability of residence times. A detailed demand allocation with stochastic demand patterns will improve water quality modelling, especially in the periphery of the drinking water distribution system.

Chapter 7

The self-cleaning velocity in practice*

ABSTRACT

The Dutch drinking water companies are constructing velocity-based self-cleaning residential drinking water distribution systems (DWDS). Field studies with particle counters have shown that these DWDS indeed do not foul. Laboratory studies have shown the settlement and resuspension of particles in water mains under constant flow conditions. However, the relationship between mains fouling and hydraulic conditions under realistic (variable) flows has not been determined.

In the presented study, the effect of variable flow velocities on particles in a real residential DWDS was studied through a detailed analysis of turbidity measurements during flushing in combination with a detailed EPANET network model. Firstly, each pipe stretch was flushed with 1.5 m/s for three turnovers and most of the pipes appeared to be clean after the first turnover. This means that it was possible to link the measured turbidity to the location in the pipe from where it was resuspended. Secondly, an all-pipes EPANET network model was filled with realistic demand patterns from the end-use model SIMDEUM; a small hydraulic time step (0.01 h) was used. This allowed for determining the maximum daily velocity occurring in each pipe stretch. The combination of the first and second step led to a graph of resuspended turbidity against the maximum daily velocity.

The study showed that in this residential DWDS there is a threshold value for the maximum velocity of 0.2 to 0.25 m/s above which the pipes remain clean. Thus, the existence of the self-cleaning velocity was further supported. The design parameters for self-cleaning networks can be fine-tuned. This study helps in understanding particle behaviour in DWDS. It may lead to more insight into what parts of the network need to be cleaned before discoloured water complaints occur.

7.1 INTRODUCTION

The goal of drinking water companies is to supply their customers with good quality drinking water 24 hours per day. With respect to water quality, the focus has for many years been on the drinking water treatment. Recently, interest in water quality in the drinking water distribution system (DWDS) has been growing. On the one hand, this is driven by customers who expect the water company to ensure the best water quality by

---

*Parts of this chapter were based on: Blokker, E. J. M., Vreeburg, J. H. G., Schaap, P. G., and van Dijk, J. C. (2010). "The self-cleaning velocity in practice." *WDSA 2010*, Tuscon, AZ.

preventing such obvious deficiencies in water quality as discolouration and (in many countries) by assuring a sufficient level of chlorine residual. On the other hand, since '9/11' there is a growing concern about (deliberate) contaminations in the DWDS. Consequently, there is an interest in the behaviour of both dissolved and particulate substances in the DWDS (Chapter 2; Powell et al. 2004).

Discolouration is the main reason for customers to complain about the water quality (Vreeburg and Boxall 2007). Vreeburg and Boxall (2007) concluded that the mechanisms leading to discolouration events are complex and poorly understood. Their basic concept of the cause of discolouration is that particles are attached by some means to the pipe wall. In normal flow the particles stay in their place and do not affect the aesthetic quality of the water. If flows are increased above normal, scouring forces and shear stress increase consequently and then the particles may be mobilised, sometimes leading to customer complaints. Vreeburg (2010) has suggested that one of the measures to reduce the discolouration risk is the prevention of the accumulating of particles in the DWDS by building self-cleaning networks. Regularly occurring high velocities and a uni-directional flow will ensure that particles are mobilised regularly and are then removed from the distribution network in small quantities through the consumers' taps. Thus, the particle accumulation will be kept within limits. In practice, the self-cleaning DWDS concept leads to a branched distribution system with sufficiently small diameters. The diameters are selected based on a design velocity of 0.4 m/s and the expected demand, which is determined with the so-called $q\sqrt{n}$ method (Vreeburg et al. 2009). This method calculates the maximum demand through a square root relationship with the number of homes on a branch.

Laboratory tests in the Netherlands (Slaats et al. 2003) have shown that sediment substitutes (sand, iron oxide and flour) of different sizes and densities are partly mobilised at velocities of 0.1 to 0.15 m/s and fully mobilised at velocities of 0.15 to 0.25 m/s. Ackers et al. (2001) found that particles with realistic diameters (45 μm) and a relatively high specific density (2600–3100 kg/m^3), started moving at flow velocities of 0.2 to 0.25 m/s. Ryan et al. (2008) have researched the velocities at which particles were mobilised; they used particles that were obtained from flushing actions throughout Australia. Mobilisation was found to occur at 0.2 to 0.3 m/s. It is unclear which flushing velocities were used to collect the particles and thus it is unclear if all particles were removed from the water mains or only the ones that could easily be mobilised. It is unclear from their study whether the material that would only be mobilised at a higher velocity was missed in their test. Husband et al. (2008) did discolouration tests in the laboratory with tap water that had a high natural iron concentration. They not only tested with constant flow velocities, but also with a dynamic flow pattern, of which the flow was changed every 15 minutes. The tests suggest that sediment accumulation and mobilisation are affected differently by dynamic flows and constant flows. At constant flows, no matter how high they are, pipes may foul. A high velocity that occurs only seldom may be able to mobilise the sediment instantaneously and keep the pipe clean.

Field measurements in the Netherlands have shown that the self-cleaning concept is feasible in real networks (Blokker et al. 2009b). Measurements with particle counters and flushing tests showed that branched systems with small diameters are cleaner (i.e. less material has accumulated) than conventional networks that were designed for fire flow demands. The field measurements also showed that the current method used to estimate the maximum flow (the $q\sqrt{n}$ method) overestimates the regularly occurring flow, meaning that the flow for which the DWDS was designed (almost) never takes place. Thus, the networks appear to be self-cleaning at velocities below 0.4 m/s. The actual velocity at which particles are mobilised is therefore smaller than the design velocity of 0.4 m/s. However, it is unclear what the value of this velocity is and how often this velocity has to occur in order for the DWDS to be self-cleaning. Also, there is a need for a more accurate method to estimate the demand.

The self-cleaning velocity is difficult to assess in a laboratory environment because it is difficult to apply realistic demand patterns as they occur in the entire DWDS. An option could be to use a stochastic demand model like the PRP model (Buchberger et al. 2003) or the end-use model SIMDEUM (Chapter 3) to simulate a demand pattern for a specific number of homes. The resulting demand pattern could be applied in a laboratory setup. Demand patterns of various number of houses should be applied to get an insight into the hydraulics of a DWDS.

An alternative is to investigate the self-cleaning velocity in a real network with real water demands and real flows. In the peripheral zones of the distribution system, flows are highly variable (i.e. show little auto-correlation and little cross-correlation). Therefore, neither flow measurements nor the standard way of hydraulic modelling (i.e. using the pumping station demand pattern) suffices to gain insight into the flow regime over time (Chapter 2). Instead, a stochastic demand model like the PRP model (Buchberger et al. 2003) or SIMDEUM (Chapter 3) should be applied. Such a model can be used to develop an alternative for the $q\sqrt{n}$ method (Buchberger et al. 2008) or allocate the stochastic water demand patterns to demand nodes in a hydraulic network model (Chapter 6). This chapter investigates whether there is a velocity above which there is no accumulation of particles in a real (Dutch) DWDS.

7.2 METHODS AND MATERIALS

7.2.1 Introduction

This chapter investigates whether there is a velocity above which there is no accumulation of particles and what the value of this self-cleaning velocity is in a real (Dutch) DWDS (Sec. 7.2.2). By using a real DWDS and a network model with realistic demand patterns, it is possible to determine the relationship between hydraulics and particle behaviour. The material that leads to discolouration problems was studied through controlled flushing tests (Sec. 7.2.3). The end-use model SIMDEUM is used to determine the velocities that occur in the DWDS during normal operation (Secs. 7.2.4 and 7.2.5). The combination of regularly occurring flow velocities and the flushed material (Sec. 7.2.6) will demonstrate the relationship between hydraulic circumstances and particle behaviour in a DWDS.

7.2.2 The network

Two areas in the town Purmerend in the Netherlands were studied. Purmerend had, from a Dutch perspective, a relatively high average of 1.5 discoloured water complaints per 1000 connections per year from 2006–2008 (Schaap and Drevijn 2009). The Resuspension Potential Method (RPM; Vreeburg et al. 2004) was used to prioritise the flushing of the network. The RPM results of February 2008 showed that the areas in Purmerend needed cleaning. The flushing was done in October 2008. An RPM was again done in November 2008 to determine the effect of the network cleaning. The average RPM of 20 locations in Purmerend before cleaning was 33.5 FTU, after cleaning it was down to 3.3. Also, the first 7 months after the network cleaning the average number of discoloured water complaints had gone down from 1.5 to 0.23 per 1000 connections per year. It was concluded that the cleaning was effective.

The Provincial Water Company North-Holland (PWN) built this network between the late 1960s and early 1980s. It is a conventional residential DWDS designed for fire flows. The total pipe length is about 22.5 km of mainly Ø100 mm asbestos cement pipes (Figure 7.1, Table 7.1). The networks have no cast iron mains. The two areas serve 4180 households - mainly single family homes, but also a few apartment buildings. The areas are fed by the same Ø500 mm AC main. This means that they have the same incoming water quality. The network was last flushed ten years before the measurements of 2008.

7.2.3 Flushing the network

The material that leads to discolouration problems is what is referred to as ‘suspendable’ particles. This material can be studied through controlled flushing tests (Vreeburg 2010).

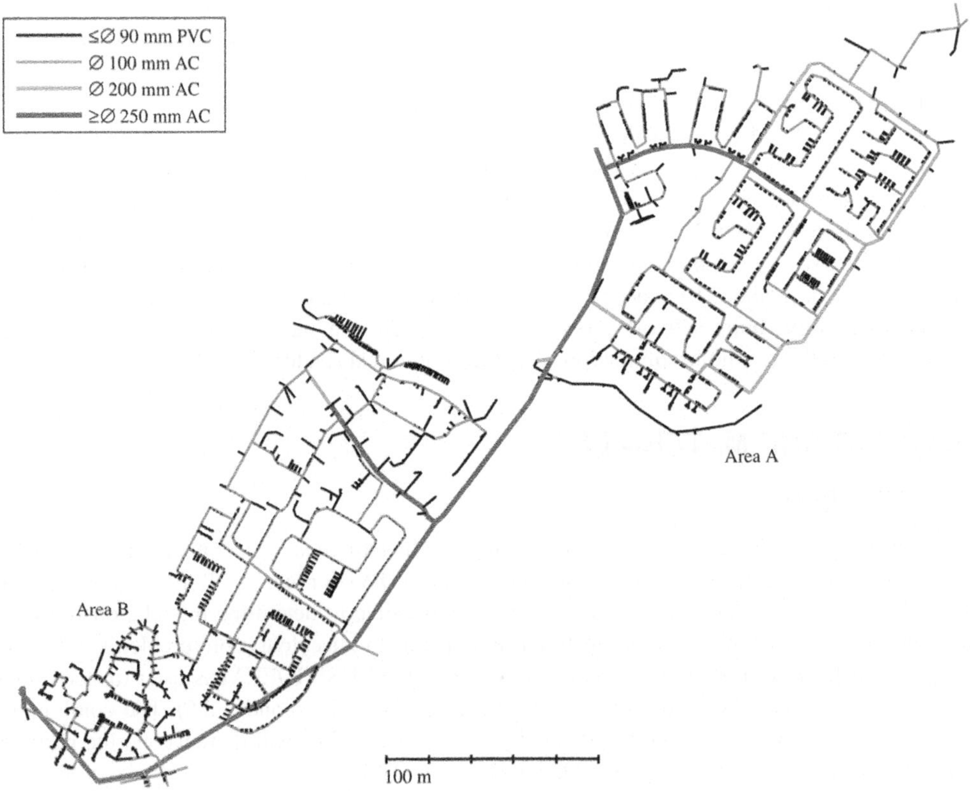

Figure 7.1 Network layout

Table 7.1 Network characteristics

	Area A (12.3 km)	Area B (10.0 km)
Ø63 PVC	5.5%	<2.0%
Ø90 PVC	<2.0%	5.3%
Ø100 AC	57.0%	54.4%
Ø110 PVC	8.7%	13.5%
Ø150 AC	<2.0%	12.6%
Ø160 PVC	4.8%	4.1%
Ø200 AC	14.0%	3.4%
Ø250 AC	3.8%	<2.0%
Ø300 AC	<2.0%	3.2%
rest	4.7%	1.6%
Total	100%	100%

The network was flushed in October 2008; this took 8 days. The flushing programme (Table 7.2) was carefully designed with the following conditions (Vreeburg 2010):

1. Work from a clear water front; flush only from a pipe that was already cleaned.
2. Use a flushing velocity of at least 1.5 m/s.
3. Flush the pipes during two to three turnovers, i.e. until the pipe volume has been renewed two to three times, or stop when the turbidity is lower than 0.6 FTU.

The flushing programme indicated the hydrants to use, which valves to open and close before and after each flushing action, and the required flow and flushing duration. During the flushing, a fraction of the flushing water (50–100 l/h) was led to a measurement unit with a 10 m (Ø 10 mm) hose. Here, the turbidity of the flushed water was measured with a Dr Lange Ultraturb SC100 with a logging frequency of once per 5 seconds. The flushing flow was recorded with a flow meter.

The analysis of the flushing results showed the cleaning of the network was successful, i.e. a low turbidity at the end of the flushing action, and that most pipe stretches were already clean after one turnover. For example in flush action 408, 94% of the total turbidity was removed in the first turnover (Figure 7.2).

In order to determine the relationship between the common hydraulic circumstances and particle accumulation and mobilisation, the measured turbidity during flushing is interpreted as the material that had accumulated before (and was mobilised during flushing) at a location that corresponds to a position within the flushed pipe length. This 'locally accumulated material' (LAM) is measured in FTU. In practice, for each flushing action, the measured turbidity of the 1st turnover was linked to the location in the stretch of pipes from which the particles originated. This was done by converting the measurement time t (s) to the flushed pipe length l (m) with the help of the flushing flow Q_f (m^3/s) and pipe diameter d (m):

$$l = \frac{t \cdot \pi(d/2)^2}{Q_f} \tag{7.1}$$

With this equation, for the first turnover, of flush action 408, for example, the turbidity over time (Figure 7.2) was converted to turbidity over the length of the pipe (Figure 7.3).

The fact that during most of the flushing actions the pipes were cleaned in the first turnover makes it plausible that flushing removed mainly loose deposits (Vreeburg and Boxall 2007). Therefore, the concept of local turbidity is a reasonable assumption. The conversion of Eq. 7.1 can only be done if the flushing flow is known (i.e. no leaking valves and the flushing was done over a single hydrant), and the start of the flushing is accurately known (which is not the case for short flushing lengths). For this research, only the flushing actions that removed >80% of the turbidity within the first turnover and for which the measured turbidity over time could be converted to LAM over the pipe length were used for further analysis (Table 7.2). Altogether, 44% of the flushed length of area A was used and 68% of flushed length of area B. In total, 9.8 km of flushed pipes was used in this study.

7.2.4 Water demand pattern generation

In this study the end-use model SIMDEUM (Chapter 3) was used to generate demand patterns. SIMDEUM is a simulation model for residential water demand patterns on a small temporal scale (1 s) and small spatial scale (individual home). For network modelling it is possible to aggregate the water demand patterns to a

Table 7.2 Overview of flushing actions

Area	Flushing action No.	Length (m)	Diameter (mm)	Flushing velocity (m/s)	Used Y/N	Reason not used
B	386	351	100, 150	1.8	N	Turbidity in 1st turnover <80% of total turbidity
	387	96.5	150	1.6	N	Turbidity in 1st turnover <80% of total turbidity
	388	118	100	2.8	N	Short length (<60s)
	389	47	100	3.2	N	Short length (<60s)
	390	323	100	1.9	Y	
	391	429.2	100	2.3	Y	
	392	16.6	96.8	3.2	N	Short length (<60s)
	393	121.2	100	2.7	N	Turbidity in 1st turnover <80% of total turbidity
	394	120	100	3.2	N	Turbidity in 1st turnover <80% of total turbidity
	395	413.9	200, 300	0.7	N	Turbidity in 1st turnover <80% of total turbidity
	396	123.9	110, 200	1.2	N	Flushed over two hydrants
	397	480.4	110, 150	1.0	Y	
	398	282.4	150	1.4	Y	
	399	234.9	100, 150	1.8	Y	
	400	479.7	150	1.3	Y	
	401	368.5	100	2.2	Y	
	402	386.7	100, 150	2.0	N	Flushed over two hydrants
	403	249.9	100	2.2	Y	
	404	63.5	96.8	3.1	N	Short length (<60s)
	405	152.2	100	1.9	Y	
	406	389.5	100	2.1	Y	
	407	184	100	2.0	Y	
	408	338	100	2.2	Y	
	409	340	100	2.1	N	Part of flushing was done with smaller flow
	410	135	100	2.2	Y	
	411	274	100	2.1	N	Turbidity in 1st turnover <80% of total turbidity
	412	64.4	100	2.8	N	Leaking valves
	413	643.6	100	1.7	Y	
	414	606	100	1.3	Y	

A	423	693	200, 250	1.8	Y	
	424	938.8	200	1.7	N	Turbidity in 1st turnover <80% of total turbidity
	425	909	200	1.8	Y	
	426	272.3	100	1.4	N	Turbidity in 1st turnover <80% of total turbidity
	427	195	100	1.7	N	Leaking valves
	428	50	100	3.2	N	Short length (<60s)
	429	328.4	100	2.2	Y	
	430				N	Data were lost
	431	222.3	100	1.8	N	Leaking valves
	432	312.2	147.6	1.4	Y	
	433	956.7	160, 110	1.8	Y	
	434	461.9	100	2.1	N	Turbidity in 1st turnover <80% of total turbidity
	435	25.6	147.6	1.1	N	Flushed over two hydrants
	436	350.1	100	2.2	N	Turbidity in 1st turnover <80% of total turbidity
	437	197.1	100	2.6	N	Turbidity in 1st turnover <80% of total turbidity
	438	357.9	100	2.8	N	Two actions flushed over same hydrant
	439	136.3	100	2.6	N	Leaking valves
	440	96.8	100	2.0	N	Leaking valves
	441	122.4	100	2.4	N	Turbidity in 1st turnover <80% of total turbidity
	442	368.4	100	2.4	N	Leaking valves
	443	277.9	100	2.5	Y	
	444	139	100	1.6	N	Leaking valves
	445	46	100	3.4	N	Short length (<60s)
	446	244.6	100	2.1	N	Turbidity in 1st turnover <80% of total turbidity
	447	456.8	100	2.2	N	Two actions flushed over same hydrant;
	448	356	100	2.1	N	Turbidity in 1st turnover <80% of total turbidity
	449	480.7	100	1.8	Y	
	450	151.4	100	2.5	N	Turbidity in 1st turnover <80% of total turbidity
	451	143.9	100	2.8	N	Turbidity in 1st turnover <80% of total turbidity
	452	513.2	100	2.1	Y	
	453	302.9	101.6	2.8	N	Turbidity in 1st turnover <80% of total turbidity

larger temporal scale (e.g. 0.01 h), or a larger spatial scale (e.g. a sum pattern of 18 homes; Figure 7.4). SIMDEUM is based on stochastic information on end uses. It predicted well the maximum velocities for 43 individual Amsterdam homes and the sum of the homes. Also, the water demand pattern over the day for the sum of the 43 homes was predicted well with SIMDEUM (Chapter 3). Another study (Chapter 6) showed that the sum of the flow pattern of 1000 homes was predicted well with SIMDEUM and that the travel times and variability in travel times was predicted well with an EPANET model with SIMDEUM demand patterns. Therefore, it was assumed that SIMDEUM would generate realistic water demand patterns for the Purmerend DWDS.

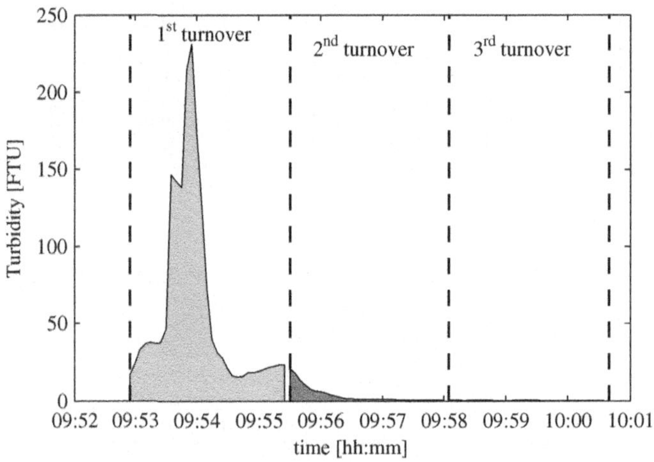

Figure 7.2 Turbidity measurement of flush action 408 at 16 October 2008

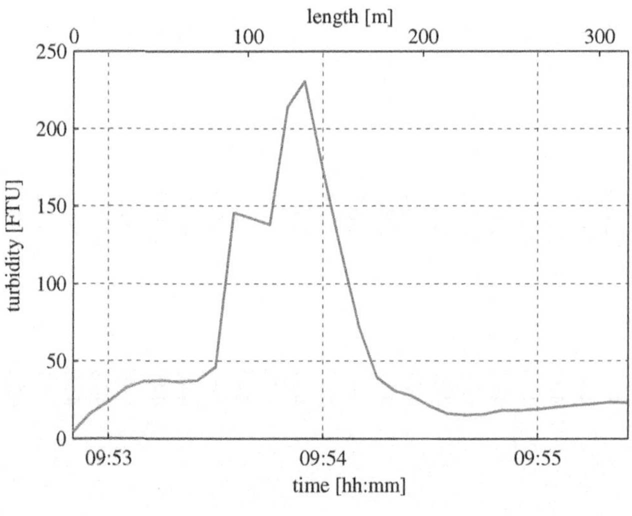

Figure 7.3 Measured turbidity, over time and length of the pipes of flushing action 408. This stretch of pipes was flushed with 2.2 m/s

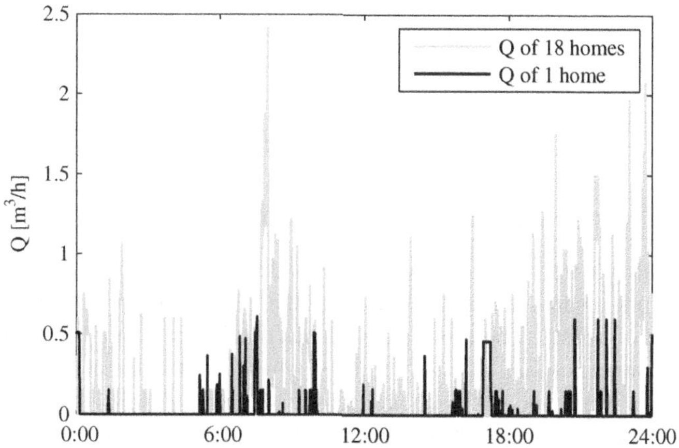

Figure 7.4 Two of the 2460 demand patterns in the network model.

SIMDEUM used generic Dutch data (Chapter 3) and some area specific data on households (Table 7.3) to generate Purmerend water demand patterns. With SIMDEUM, 2500 different homes were simulated, with different household sizes, different sleep-wake rhythms of the residents, and different water-using appliances. For each home, 10 different patterns were generated. These patterns were applied to areas A and B (Sec. 7.2.5).

Table 7.3 Specific Purmerend 2008 (CBS) input data into SIMDEUM

		Purmerend
House-holds	One person households	31.0%
	Households without children	33.2%
	Households with children	35.8%
	Average household size	2.3
Age distribution	0 to 12 years old	15.5%
	13 to 20 years old	10.0%
	21 to 64 years old	60.4%
	65 years and older	14.1%

In a separate simulation run, 500 different homes were simulated with SIMDEUM, with different household sizes, different sleep-wake rhythms of the residents and different water-using appliances. For each home, 30 different patterns were generated. Then, patterns were randomly summed for a 'street' of 2, 5, 10, 20, 50, 100, 150, 200 and 250 homes, each with 500 patterns. Of these patterns, the maximum flows were determined. The median and 99th percentile of these maximum flows, Q_1 and Q_2 (in L/s) respectively (in Eqs. 7.2 and 7.3), were then determined and a relationship between these flows and the number of supplied homes was fitted on the data; the fitted relationship is similar to the one from

Buchberger et al. (2008).

$$Q_1 = 0.157 + 0.073\sqrt{n} + 0.009n \quad (7.2)$$

$$Q_2 = 0.242 + 0.139\sqrt{n} + 0.008n \quad (7.3)$$

7.2.5 The hydraulic network model and demand allocation

In order to get detailed insight into the local hydraulic conditions in the network during normal operation, there is a need for a network model with many short pipes that all experience a different flow pattern because they serve a different set of homes. Therefore, an all-pipes network was used that includes the connection pipes. Moreover, each connection has its own unique stochastic water demand pattern (Sec. 7.2.4).

A detailed hydraulic network model of each area was made in EPANET (Rossman 2000). The hydraulic models of both areas consist of approximately 3000 pipes and 3000 nodes. The hydraulic models have demand nodes for each physical connection; area A and B have 1133 and 1327 demand nodes, respectively. Each demand node was assigned a unique stochastic SIMDEUM demand pattern. The nodes connecting a single family home were assigned a demand pattern of one home; the nodes connecting an apartment building were assigned a demand pattern of the sum of the appropriate number of single home demand patterns; Figure 7.4 shows an example for a single home and the sum of 18 homes. The model time scale must be sufficiently small. Based on the expected flows within the peripheral zone of the DWDS, where most pipes supply to less than 200 homes, a time step of 1 minute is short enough to capture the maximum flow velocities (Chapter 2). The demand pattern time step in this study was set to 0.01 h (36 s).

As velocities change constantly, both in reality and in the model with SIMDEUM demand patterns, a statistical approach is required to determine the normal hydraulic circumstances. The networks in the two areas are strongly looped (Figure 7.1). Using a different random set of water demand patterns does have an effect on, for example, the maximum velocity per pipe. Therefore, ten sets of random water demand patterns were used for this study. All ten sets of water demand patterns are representative of average days; a maximum water demand situation due to extra outdoor water use was not enforced. It is unclear if ten sets of water demand patterns are sufficient to get good insight into the variability of the maximum flow velocities, but it was expected to be enough to find the median value.

Ten copies of an EPANET.inp file were created and then unique stochastic demand patterns were imported. To the list of patterns, demand pattern IDs were added plus 2400 demand multipliers (pattern time step was 0.01 h) that were generated by SIMDEUM. The average of the SIMDEUM demand multipliers is not equal to 1 but to the actual average residential demand (L/s). Each junction that represents residential demand was assigned a base demand of 1 and a unique demand pattern ID.

7.2.6 Determining the relationship between hydraulics and settled sediment

As velocities change constantly, a statistical approach is required to determine the velocity that is characteristic for particle accumulation and mobilisation. The hypothesis of self-cleaning networks states that velocities above a certain threshold value (the self-cleaning velocity) will keep the pipes clean if they occur regularly. In case "regularly" means once per day, the maximum daily velocity is of interest. In case "regularly" means 15 minutes per day, the 99th percentile of the velocity is of interest. It is unknown how often the self-cleaning velocity must occur. However, per pipe, the maximum velocity and the more often occurring velocities are not independent variables. With $R^2 = 0.90$ and $R^2 = 0.96$ between the maximum velocity and the 50th and 95th percentile, respectively, of the velocity (Figure 7.5), there is a strong correlation. This means that the choice of the percentile of daily velocities is almost arbitrary. In this study, the maximum daily velocity is used as a measure for the hydraulic circumstances in a pipe that are related to sediment mobilisation. From the ten different EPANET simulations over 24 h, the maximum velocities in each pipe segment were extracted. For the pipes of flush action 408, the maximum velocity of one EPANET simulation is shown in Figure 7.7.

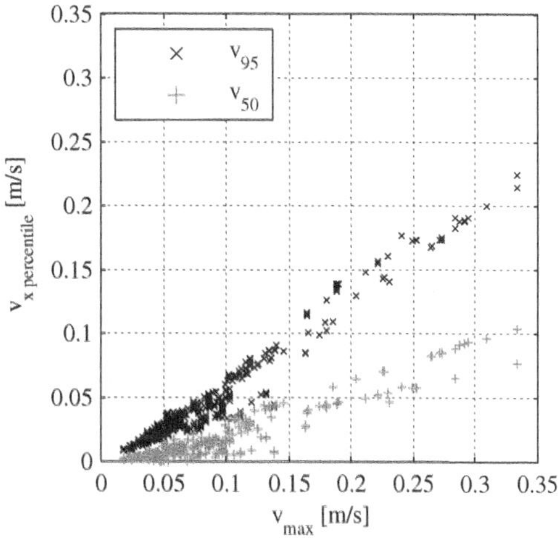

Figure 7.5 The relationship between maximum velocity and 50th and 95th percentile of simulated velocity of one day in flushed pipes of area A

For each pipe in the hydraulic network model, the maximum velocity v_{max} was determined, changing with every house connection on the pipe stretch (the left hand graph in Figure 7.6). This was done for 10 simulations; the minimum, median and maximum v_{max} were extracted. For each pipe in the hydraulic network model, the measured LAM in the centre of the pipe segments was determined (the purple coloured lines in Figure 7.6). Next, for each pipe, v_{max} and the LAM were paired (the crosses and circles in Figure 7.7). With an average pipe length of 4 m, this approach led to 1169 combinations of maximum velocity (m/s) and LAM (FTU).

Figure 7.8 shows the measured LAM and the minimum, median and maximum of v_{max} from 10 simulation runs. Figure 7.9 shows the LAM and median v_{max} together with a histogram showing the number of occurrences per v_{max}. Most v_{max} - LAM combinations are found in the low velocities, 61 combinations are found for $v_{max} \geq 0.2$ m/s. A different way of representation is given by Figure 7.10; it shows the cumulative frequency distribution (CFD) of the number of occurrences and the sum of LAM over v_{max}. At velocities below 0.07 m/s the CFD of both lines are equal, after that the lines start to deviate.

7.3 RESULTS AND DISCUSSION

7.3.1 Interpretation of results

Figure 7.8 shows that above a certain threshold value of v_{max}, the LAM is low; the threshold value is 0.2 to 0.25 m/s. Figure 7.9 and Figure 7.10 are included to support the fact that the low LAM values at higher velocities are not just an artefact of the relatively low number of data at velocities above 0.2 m/s. There are 61 occurrences (5-6% of the data) at a velocity above 0.2 m/s and the CFD of the LAM is almost 1 at 0.2 m/s and above (Figure 7.10). To further substantiate this, a statistical analysis was done. The set of LAM data (without the pairing with v_{max}) can be described with a lognormal distribution. The parameters of the lognormal distribution (M, S) were estimated from two datasets: the LAM data at v_{max} below 0.2 m/s and the LAM data at v_{max} of 0.2 m/s and higher. The parameters M and S are different, at the 95% significance level, for the two datasets. This suggests that above 0.2 m/s there is a different mechanism for particle accumulation.

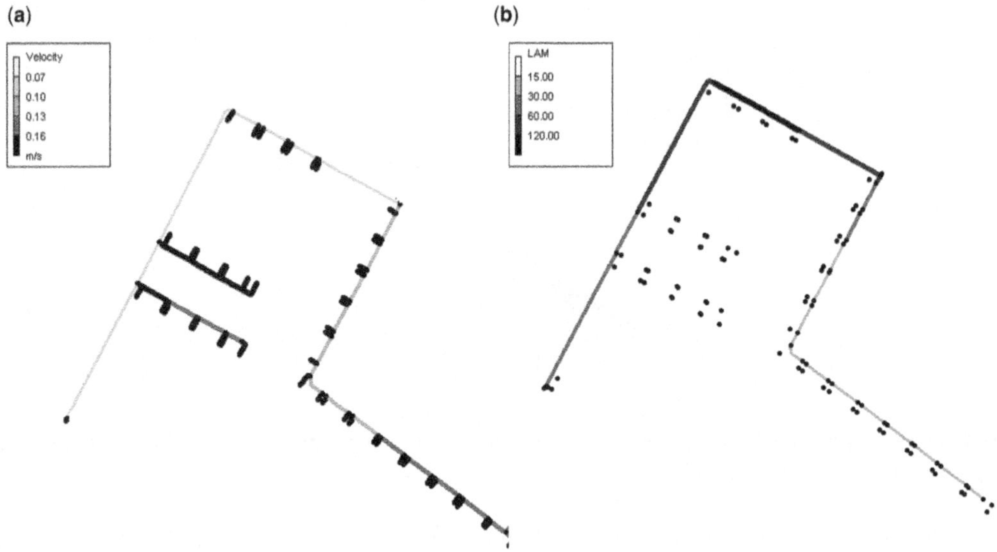

Figure 7.6 a) Maximum velocity during normal operation (simulated) and b) LAM, over time and length, on the same stretch of pipes of flushing action 408. The none coloured lines were not flushed, no LAM was determined

The LAM values above 400 FTU at 0.13 m/s are solely determined by the results of one flushing action (flushing action 407, the grey dots in Figure 7.8). This peak stands out from the rest of the values. It might be an anomaly. Maybe, hydrants have been used over the past years; or a valve that was supposed to be open may have been closed at some point. Since any historic network operations are unknown, it is unclear if this peak is an anomaly or due to normal variance.

Figure 7.8 shows that most of the particles will settle at a maximum velocity of 0.07 to 0.10 m/s. Figure 7.10 supports this; it shows that the CFD of the LAM has a larger slope than the count of v_{max} between 0.07 and 0.1 m/s. The maximum turbidity at a velocity above 0 m/s suggests that the rate at which particles deposit is related to the horizontal velocity of the drinking water, or the shear stress on the pipe wall. One possible explanation is that at very low velocities (< 0.05 m/s) only small amounts of water are flowing through and thus only small amounts of particles are brought to these locations. Another possible explanation is that the particles have settled in a prior part of the network. A third explanation is that different deposition mechanisms are at work. Ryan et al. (2008) suggest gravitational settling below a velocity of 0.07 m/s, and above that a 'wall deposition' mechanism. Vreeburg and

Boxall (2007) claim to have observed gravitational settling (sediment settled at the bottom of the pipe) at 0.06 m/s and turbophoresis (particles stuck at the total circumference of the pipe) with a high concentration of FeCl and a flow velocity of 0.14 m/s.

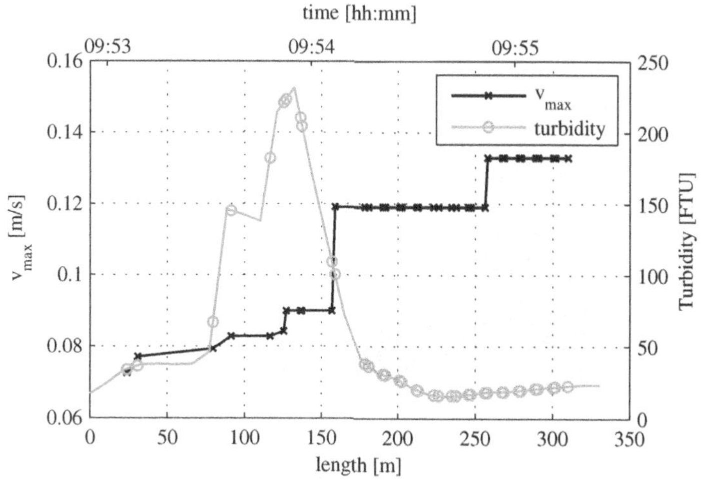

Figure 7.7 Maximum velocity during normal operation (simulated) and measured turbidity, over time and length, on the same stretch of pipes of flushing action 408

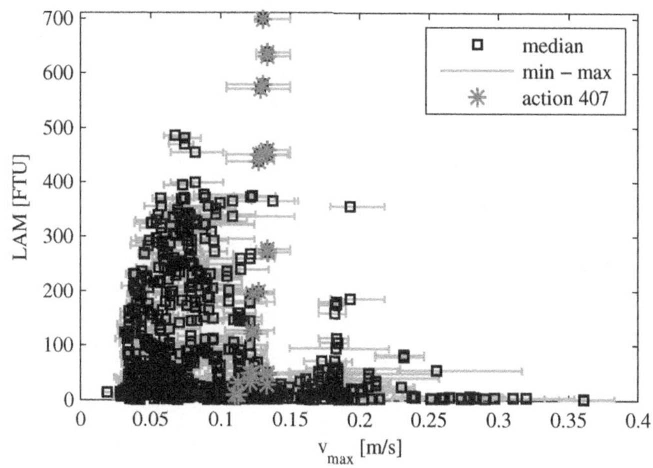

Figure 7.8 Locally accumulated material [FTU] versus maximum velocity [m/s] in areas A and B. The median, minimum and maximum values of ten simulated maximum velocities are shown

Figure 7.8 shows that below 0.2 m/s the turbidity can be high, but it can also be low. This can be explained by the fact that particles will not deposit instantly, but rather gradually. At a location with a low velocity, the particles start moving towards the pipe wall. It may take a few 100 metres before the particles actually deposit. To predict where in the network most particles will accumulate, a hydraulic network model plus a particle accumulation and mobilisation model is required.

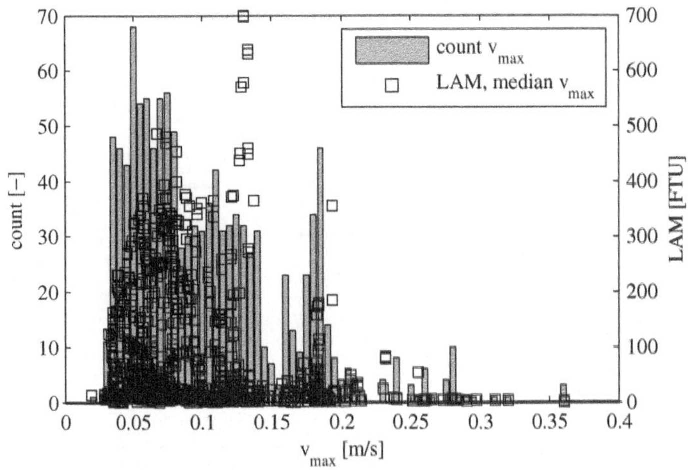

Figure 7.9 On the left axis, the number of occurrences of v_{max} at a bin size of 0.005 m/s. On the right axis, LAM versus the median of v_{max} in areas A and B

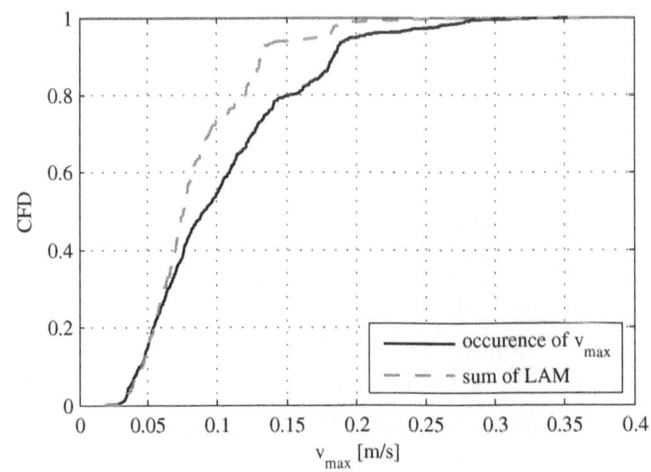

Figure 7.10 Cumulative frequency distribution (CFD) of number of occurrences of (the median of) v_{max} and the sum of the LAM in areas A and B

7.3.2 Theory of self-cleaning

The current field experiments show that a maximum velocity of 0.2 to 0.25 m/s that occurs once per two days (median of the maximum velocities) is enough to keep the pipes clean. This further supports the theory of self-cleaning, which says that a certain velocity will mobilise the loose deposits instantaneously and that, if this velocity is reached regularly, the particles will not accumulate in the pipes.

The threshold value of 0.2 to 0.25 m/s corresponds to the mobilisation velocities that were found in several laboratory experiments (Ackers et al. 2001; Ryan et al. 2008; Slaats et al. 2003). A self-cleaning velocity of 0.25 m/s corresponds to a shear stress of about 0.2 N/m^2 (using the Darcy Weisbach equation and a wall roughness of 0.2 mm for AC pipes). This corresponds to the conditioning shear stress above which the discolouration complaint level in two UK water companies was found to decrease (Cook et al. 2009). However, the velocity of 0.2 to 0.25 m/s may be specific for the material type in this network. According to Berlamont (in Vreeburg 2010), the velocity for the mobilisation of material in sewer pipes is related to particle characteristics, such as size and density.

The cohesive layer theory (Boxall et al. 2001; Vreeburg and Boxall 2007) is based on the concept that discolouration material is held in stable cohesive layers attached to the pipe walls of distribution systems and that these layers are conditioned by the usual daily hydraulic regime within the system. The occurrence of disequilibria hydraulic conditions (burst, increased demand etc.) may expose the layers to shear stress in excess of their conditioned cohesive strength and lead to a mobilisation of the cohesive layers, resulting in a discolouration event. The theory of self-cleaning is not in contradiction with this theory. In the periphery of the DWDS with low average velocities and hours of stagnant water, the conditioning shear stresses are low. During the daily maximum flows, the cohesive layers are mobilised. However, this will not lead to discolouration events, as only a small amount of particles are mobilised. Accumulation of particles can thus be prevented, and self-cleaning pipes are feasible.

As a characteristic value of hydraulic circumstances that lead to mobilisation, the median of v_{max} was used. In the network being studied, there is a strong correlation between maximum velocity and, for example, the 95th and 50th percentile velocities (Figure 7.5). Due to limited variability in pipe material and pipe diameters, there is also a strong correlation between maximum velocity and maximum shear stress. 70% of the data was collected on Ø100 mm AC pipes; 15% on Ø150 mm AC pipes and 5% on Ø200 mm AC pipes. Respective velocities for a shear stress of 0.2 N/m^2 (using the Darcy Weisbach equation and a wall roughness of 0.2 mm) are 0.23 m/s, 0.245 m/s and 0.255 m/s. The differences are too small to determine if the critical parameter is the maximum velocity, the 99th percentile of the daily velocities or the maximum shear stress (as used in the cohesive layer theory). In a large Ø600 mm AC transport main, a shear stress of 0.2 N/m^2 corresponds to a slightly higher velocity (0.29 m/s). Repeating the tests in large mains may give a more definitive critical parameter. As velocity is an intuitive value and the shear stress needs to be calculated, the velocity is preferred in practice. As a result, the maximum velocity is considered to be the actual self-cleaning velocity.

7.3.3 Practical implications

Pipe stretches with flow velocities below the threshold value of 0.2 to 0.25 m/s may foul and lead to discoloured water complaints. This means that with this threshold value, water companies can determine where they need to clean the network. This study may provide insight into a threshold value below which there is a minimum risk of discoloured water complaints. A prerequisite is an analysis of the relationship between the amount of accumulated material and the number of discoloured water complaints. Also, the threshold value can be used in the design of self-cleaning networks.

7.3.4 Design implications for self-cleaning networks

The self-cleaning design principles determine a maximum pipe diameter (Buchberger et al. 2008; Vreeburg et al. 2009) based on the $q\sqrt{n}$-method and a design velocity of 0.4 m/s. Blokker et al. (2009b) showed that networks that were self-cleaning were designed with the following equation for the q$\sqrt{}$n-method (using 15 for the so-called number of tap units).

$$Q_3 = 0.083\sqrt{15n} = 0.321\sqrt{n} \tag{7.4}$$

A minimum pipe diameter is often determined with the same equation for the flow and a maximum design velocity of 1.5 m/s. SIMDEUM was used to determine the flow which occurs every other day and the flow which occurs once per 100 days (Sec. 7.2.4). The maximum pipe diameter could then be determined with Q_1 (the median of Q_{max}, Eq. 7.2) and a velocity of 0.2 m/s, and the minimum pipe diameter can be determined with Q_2 (the 99th percentile of Q_{max}, Eq. 7.3) and a velocity of 1.0 m/s.

The resulting minimum and maximum pipe diameters are shown in Figure 7.11; available pipe diameters are plotted on the vertical axis. Up to 50 homes, there is only a small difference between the q$\sqrt{}$n and the SIMDEUM approach. Beyond that, the SIMDEUM approach leads to larger pipe diameters than the q$\sqrt{}$n-method. As pipe diameters are only available in certain sizes, this means that the SIMDEUM approach would use a PVC 110 for 82 homes and more, where the q$\sqrt{}$n-method would use it for 100 homes and more. With a maximum section size of 60 to 250 homes for new networks (Mesman and Meerkerk 2010), both approaches will lead to the selection of roughly the same pipe diameters. This means that the q$\sqrt{}$n approach with 15 tap units will lead to self-cleaning networks. Most Dutch water companies use a higher value for the number of tap units (Mesman and Meerkerk 2010). Their networks will thus experience velocities that are too low to lead to a self-cleaning network.

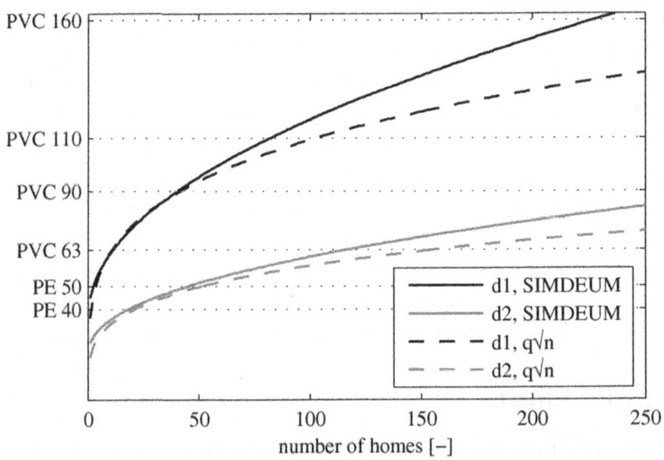

Figure 7.11 The maximum (d1) and minimum internal pipe diameters (d2) determined with two different design approaches. SIMDEUM stands for the flow Q_1 (Eq. 7.2) plus a velocity of 0.2 m/s and the flow Q_2 (Eq. 7.3) plus a velocity of 1.0 m/s to determine d1 and d2, respectively. q$\sqrt{}$n stands for the flow Q_3 (Eq. 7.4) plus a velocity of 0.4 m/s and a velocity of 1.5 m/s to determine d1 and d2, respectively.

7.4 CONCLUSIONS

This presented method of analysing flushing results together with a detailed network model with realistic demand patterns provides insight into the particle behaviour under normal flow conditions.

The flushing directive of using a clear water front, flushing uni-directionally with 1.5 m/s for three turnovers, works well for cleaning mains. After flushing, the discoloured water complaint level in Purmerend went down from 1.5 to less than 0.3 per 1000 connections per year and the average Resuspension Potential value went down from 33.5 to 3.3 FTU. In most pipe stretches, more than 80% of the total turbidity was removed in the first turnover. This enabled the assignment of measured turbidity to a location within the pipe stretch.

By analysing flushing results in combination with a network model with realistic demand patterns, it was shown that there is a velocity above which a pipe does not foul. It thus further supports the theory of the self-cleaning velocity. Pipe stretches that experience a maximum velocity of 0.2 to 0.25 m/s once per two days stay clean. Pipe stretches that experience a maximum velocity below this threshold may foul. The amount of accumulated particles depends on the hydraulic circumstances preceding to those pipes and possibly the particle characteristics.

Chapter 8

General discussion and outlook

8.1 INTRODUCTION

This chapter discusses several aspects of the work that was done; it ties the previous chapters together.

First of all, the development of the end-use model SIMDEUM and the approach of modelling the end user are discussed. This approach is possible because people as a group show predictable behaviour. The level of predictability leads to good results in water demand modelling. Although the description of human behaviour is extensive, building an end-use model is not very difficult at all.

Secondly, the application SIMDEUM in case studies and, thus, more detailed water demand patterns in network solvers is discussed. This includes a section on measurements in real drinking water distribution systems. A number of tests were conducted in actual DWDS and something always went wrong. The reader may learn from this.

Thirdly, the added value of SIMDEUM is evaluated. The added value is related to water quality modelling.

Finally, future application areas are considered. These relate both to ready-to-implement applications and further research.

8.2 CONSTRUCTING THE END-USE MODEL SIMDEUM

8.2.1 The approach of modelling the end user

Residential water use over the day can be considered a stochastic process. It is, however, not completely random. Water use is usually low during the night and high in the morning and evening. One approach towards modelling water demand is to measure water demand patterns and then try to describe the water demand with relatively simple probability distributions. Another approach is to try and understand the behaviour with respect to water use and find probability distributions to describe that. The first approach is that of the PRP method; the latter, that of SIMDEUM (Chapter 5).

In the end-use modelling approach the diurnal water demand pattern is explained. The relative height of the peaks in water use, and at what time they occur, is related to people being at home and them being active (i.e. awake). The height and duration of individual water demand 'pulses' is related to the type of end use, e.g. filling a toilet cistern or taking a shower, and is thus determined by the end-use characteristics, e.g. the volume of the toilet cistern or the capacity of a shower head.

The approach of modelling the end user is possible because people are creatures of habit. People tend to get out of bed at approximately the same time every work day, leave the house, return home and go to bed at the same time. Also, in clusters, people behave alike. Children in the Netherlands start school at around 8:30 and finish school at 15:00. This influences their sleep-wake rhythms and those of their parents.

Different countries show different typical demand patterns. The Dutch residential water demand pattern has, on work days, a very distinct peak in the morning and a lower peak in the evening (Figure 8.1). The residential water demand pattern of Milford, OH, USA, has two peaks (one in the morning, one in the afternoon) that are comparable in intensity (Figure 8.1). The residential water demand pattern of Piedimonte San Germano, Italy, has one long enduring peak in the morning (Figure 8.1); the night use is high because of leakage (Tricarico 2005). The relative height of the peaks, and at what time they occur, is related to users being at home and active; this appears to be very much related to culture. In Milford, people do not get up all at the same time; some of them considerably earlier than is customary in the Netherlands or Italy, for example. This is supported by the ATUS data (Chapter 5).

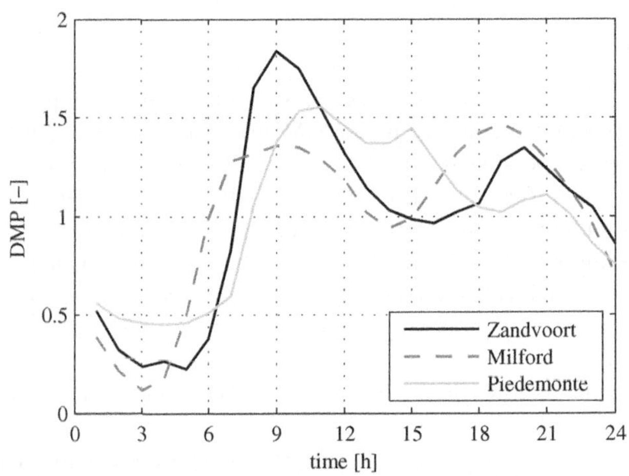

Figure 8.1 Residential demand multiplier pattern (DMP) of Zandvoort, the Netherlands, in Sept–Oct 2008 (Chapter 6), Milford, Ohio, USA in May–June 1997 (Buchberger et al. 2003; Chapter 5) and Piedimonte San Germano, Italy, in Dec 2001 (Tricarico 2005)

8.2.2 Advantage of end-use modelling

The advantage of the end-use approach is that the effect on the water demand of using different appliances or having a different culture can easily be determined. There is no need to do flow measurements; information on time use and water-using appliances is often available. This is especially convenient in the design stage of a network and in scenario studies, where doing flow measurements is impossible.

With this approach, it is also possible to simulate the water demand by non-residential users (Chapter 4). These may have different appliances than households, have more end users present, and the active time is determined by other factors. For example, offices have industrial coffee machines and only experience water use during office hours. That means that water companies can have different water demand patterns for residential and industrial areas and not use the same pumping station curve for the whole network.

The approach of simulating demand patterns of drinking water can also be used for simulating electricity or gas demand patterns. There is a limited model available for residential energy use over the day (Paauw et al. 2009).

8.2.3 An undemanding model

The approach of modelling the end user is possible because people, as a group, show predictable behaviour. The level of predictability leads to good results in water demand modelling. Although the description of human behaviour is extensive, building an end-use model is not very difficult at all. It was shown that the main drivers for water use are related to people's sleep-wake rhythms and their being at home. The particular water use is determined by the end-use characteristics, e.g. the capacity of the toilet cistern or the flow of a shower head.

The data for the end-use model were collected for the Netherlands. It is straightforward to use SIMDEUM also outside of the Netherlands. As input data, no flow measurements are required and no calibration on demands is needed. For the input data, much of the Dutch data can be re-used and country specific data can often be found in the literature. The required input data are partly generic, e.g. the number of times people use the toilet will not be different in the Netherlands and in Portugal. Also, other countries may have the same appliances that are used in the Netherlands. The input data are partly specific, e.g. some washing machine types are only sold in Europe. Also, people in different cultures have different diurnal patterns. The Centre for Time Use Research (ref) has links to data from time-use surveys all over the world. The literature on water use, often aimed at coarse water demand management, is available for many countries.

The application in the study of water use in Milford, Ohio, showed that the data were collected within one day (Chapter 5).

8.2.4 Pressure driven demand

Pressure influences demand to a certain extent. When there is no pressure, there is no water use. Water demand may shift to the time when water is available again. People still want to take a shower and flush the toilet. When pressure is low, various water-using appliances are affected differently. The toilet cistern will be filled with a lower flow, but the total demand remains the same. This applies to all water use that is related to a certain volume, such as washing machines and dishwashers. The water use at a running tap will decrease in flow with low pressure, and potentially increase in duration. Washing hands may take longer with low flows, but the total water use for washing hands at a low flow washstand may be less than at a high flow washstand. People may decide not to take a shower at times of low pressure and shift their shower to a later time when pressure is sufficient again. In the case of very high pressures, people tend to pinch the flow of the tap to prevent spattering. High pressures are not likely to lead to more water use. The per capita water use of consumers living in high pressured DWDS areas is not higher than of consumers living in normal pressured DWDS areas.

The input parameters of SIMDEUM (Chapter 3) were determined in normal pressure situations. They can be adjusted to account for the influence of pressure in the DWDS. In scenario studies, especially in low pressure zones, this is something to consider.

Since leakage is low in the Netherlands, both indoors (Foekema et al. 2008) and in the DWDS (Beuken et al. 2006; Geudens 2008), leakage was not included in SIMDEUM. For other countries, leakage at the cumstomer premises can be incorporated in the end-use model as a low continuous pattern of drops or continuous flow, or it can be included as a pressure-driven demand. With a droplet size of circa 0.35 ml, a leaking tap of 0.1 to 1.0 L/h can be modelled as one drop of 0.35 ml in one second and then several seconds of no flow. Leaking toilets or plumbing can be modelled with a continuous flow of circa 1 to 10 L/h. The modelling of leaking distribution mains in hydraulic network solvers is a different issue, and object of many studies.

8.3 CASE STUDIES – MEASUREMENT ISSUES

Although this book is mainly concerned with modelling, quite a large number of measurements were performed as well. Measuring is not easy; many things can go wrong. There is usually only one chance to do it right for reasons of time consumption and cost. In order to learn from our mistakes, what follows is an abstract of what went wrong or did not go as planned.

For the purpose of model validation, several measurements were done in real networks. In 2006 a tracer test was done in a small network; unfortunately, the test gave no results other than how **not** to perform such a test. The NaCl dosage setup did not work; mixing was not reached because the salt sank directly to the bottom and no rise in electrical conductivity was measured. A static mixer was installed in the succeeding tests. In 2007 a successful test was performed in the same network (Chapter 6). We learnt a lot from these tests and performed a similar tracer test in 2008 in a network ten times the size of the first one (Chapter 6).

Choosing the correct setting for flow measurements is prone to errors. Instead of average flow, sometimes the instantaneous flow was logged. Choosing the proper logging frequency with the suitable pulse resolution is important. A good feel for expected flow values is indispensable; a 100 L/pulse with a logging frequency of 1 minute in an apartment building of only 54 flats is not useful; low continuous flows were interpreted as separate short pulses.

After the measurements are done, it is time for interpretation of the data. Often, this demonstrates that information is lacking. This is especially true for measurements that were done years earlier and for different purposes than the present analysis. Most often these data were not suitable for new purposes. The measurement setup description was not complete: for example, extra measurements were made without noting when and how much additional water was extracted. Statistical analyses of the measurement data sometimes showed that too few data were collected for a significant outcome. Unfortunately, it was not possible to extend the measurement period as the setup was already dismantled.

In cases when differences between two measurements were found, it was unclear which one was correct. A difference between household water meter readings and the flow meter in Franeker was found; the water meters measured about 15% more water use than the flow meter. A difference between the bulk water meter and flow meter in Zandvoort was found; the bulk water meter did not measure night use and presumably logged instantaneous flow rather than average flow, and the flow meter may have used the wrong diameter to compute flow from the measured flow velocity.

The electrical conductivity (EC) measurements in Benthuizen (Chapter 6) showed that the EC after the first few NaCl pulses was not constant during the dosage nor was it constant between two dosages. The variability of the EC did not negatively influence the measurements, but could not be explained either. Was there a leaking valve that resulted in NaCl leaking back into the measurement area?

The additional water extraction in Benthuizen (to limit longitudinal spreading) was set at 400 L/h. After six days, the additional extraction had dropped to approximately 280 L/h. From the flow measurements, it was most likely that the flow gradually decreased, but this remained an estimate.

During the summer of 2006, the water demand was higher than average. The water companies explained this by reminding of the extremely high temperatures during that period. External factors to explain deviations are not always that obvious. Nor is it easy to prove from the original measurements that this external factor is the cause. Extra data, therefore, is needed.

A large part of the measurement data that were used were collected by other parties. For example, the survey data on water use and time use were collected by Vewin and SCP, respectively; the flow measurements of Milford, OH, were provided by the University of Cincinnati, and flow measurements of the (non-)residential users were supplied by several Dutch water companies. The Provincial Water Company North-Holland PWN provided detailed turbidity data from flushing networks. The data were critically analysed and found to be fit for the purposes of the book, although sometimes there was a need for more detail in the data, or not all meta-data were available.

8.4 CASE STUDIES – NETWORK SOLVER CONSIDERATIONS

In the case studies of Chapters 6 and 7, very detailed demand patterns were applied in hydraulic network models. The simplifications of standard hydraulic network solvers need to be reconsidered for these circumstances.

In traditional hydraulic models, the time step that is used most often is 1 h. For water quality modelling, a shorter time step may be more suitable. The most suitable time step depends on the application, e.g. determining chlorine residual or particles resuspension. It also depends on the location in the network, e.g. in a feeding main where flows are not very variable or in an end branch where water is stagnant during most of the day. From using a short time step, it can be established if temporal aggregation is possible without losing too much information. In modelling travel time in the networks discussed in Chapter 6, a time step of 5 or 15 minutes appeared to be sufficient. In modelling maximum velocities in the networks of Chapter 7, a time step of maximum 1 minute was required.

In current hydraulic network modelling, transients are neglected. The applications discussed in this book are water quality modelling related to travel time, temperature and mobilisation of particles. The most suitable time step for these applications is 1 min or longer. At this time scale, transients are not important.

The measurements of tracer pulses in Benthuizen were designed to limit dispersion in the test case. The measurements of tracer pulses in Zandvoort-emphasised dispersion is an issue in real distribution networks (Chapter 6). In typical residential DWDS such as Purmerend (Sec. 8.5.1), the flows are mainly laminar and a network solver that considers dispersion should be applied when dissolved substances are studied (Lee 2004; Li 2006).

Several authors have shown that full mixing at cross-junctions is an invalid assumption (Austin et al. 2008; Ho et al. 2007; Romero-Gomez et al. 2008a). They have tested with constant flows, or constant ratios between flows in the incoming pipes (south and west; Figure 8.2). In reality, flows do not show 100% cross-correlation, and thus the (possibly contaminated) water coming from the west may at one time step flow to the north and at another time step flow to the east. In a water quality model with a hydraulic time step of 1 h, this may mean that full mixing can be assumed after all. In the Netherlands, cross-junctions are not used, so this hypothesis has not been tested.

The use of stochastic demand patterns means that the result of one network calculation is only one possible outcome. The same calculation with a different set of demand patterns may lead to different results. In Chapter 6, ten different model runs were used. The resulting travel times varied as much as 40% of the average value. This is much closer to the real variation in travel time than the conventional top-down approach showed. In Chapter 7, also ten different model runs were used. The resulting maximum velocities varied 20% (on average) to 77% (maximum) compared to the average result. How many model runs are required in order to draw conclusions with a certain confidence interval is unknown.

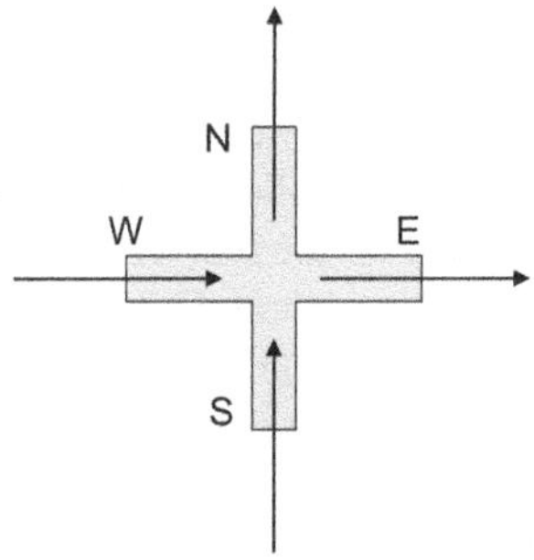

Figure 8.2 Pipe cross-junction with incoming flow at south and west side and outgoing flow at north and east side

Most water quality processes are accumulation processes. Therefore, the hydraulic conditions as they occur over the year are important. This means that mainly average flows with regular variability are required. To get the full range of possibilities (i.e. also the incidental maximum flows), it may be smarter to specifically run the model with extreme situations, instead of running the model with multiple sets of random demands. To get a good representation of the extremes in the model, the best approach is to determine what causes the extremes and put that into the end-use model. For example, the maximum annual water use in the Netherlands is caused by outdoor water use in the summer when people water their gardens and fill children's baths. Low water use may happen during holidays when people are not at home.

8.5 EVALUATION OF ADDED VALUE OF SIMDEUM

8.5.1 Effects of detailed demand model in DWDS network model

There are numerous water quality aspects that may change during distribution, for example contamination propagation, corrosion, sedimentation and resuspension, coagulation, flocculation, precipitation, disinfectant decay, disinfection by-product formation, bacterial re-growth, biofilm formation, nitrification, taste and odour. These aspects are influenced by the fact that dissolved or particulate substances move with the water (advection, dispersion, mixing at junctions, etc.) and that they can react with the wall or other substances. The reactions may be influenced by contact time (travel time), temperature (which is also influenced by travel time) and shear stress or flow velocity. This means that the hydraulic circumstances are very important in water quality processes. The choice of a water demand model has a large effect. The chosen time scale will affect the probability of finding stagnant and laminar flows, and thus affects the dispersion model. The chosen spatial scale will affect the probability of finding flow direction reversals and the cross-correlation of flows. Different sets of demand patterns (from different simulation runs) may give insight into the variability of travel times and flow velocities.

In this section, the effect of a detailed demand model on the parameters affecting the water quality modelling is investigated. In Chapter 2, the effect of temporal and spatial aggregations of realistic water demand patterns was shown for a limited number of homes, and only in end branches, not in a looped network. To study the same aspects in a real network model, the area A in Purmerend (Figure 8.3, see also Chapter 7) was studied in more detail. A detailed demand allocation, in this case, means a short time scale, and each connection has its own unique demand pattern. The effect of the stochastic nature of the demand model was not investigated to the fullest in this case; only one set of demand patterns was applied.

Figure 8.3 DWDS of area A in Purmerend. The colours indicate diameter

It is not possible to measure flows in an entire network. Instead, SIMDEUM was used to get a view of the actual demands in a network. In previous chapters, SIMDEUM was validated and can be assumed to give realistic demands. Thus, hydraulic network simulations with SIMDEUM demand patterns enable a quantification of the error that a traditional hydraulic network model causes.

The Purmerend area A was modelled in EPANET (Figure 8.3) with a time step of 0.01 h and the bottom-up approach of demand allocation (BU model); the results of one set of demand patterns are shown and discussed here. This network was also modelled with a time step of 5 min and the top-down approach of demand allocation (TD model); using the sum pattern of the bottom-up approach as DMP. The results of the top-down model are not shown here graphically, but they are being discussed.

Figure 8.4 shows the cross-correlation between the incoming flow and the flow in all pipes for the BU model. The larger pipes have a high cross-correlation; 28% of the distribution pipes (pipes with diameter >60 mm, weighted over the pipe length) have a cross-correlation of 0.8 or more. The smaller pipes, especially the ones that feed a limited number of homes on a branch, show a small (<0.5) cross-correlation; 35% of the distribution pipes have a cross-correlation of 0.5 or less. The TD model has a cross-correlation of 1 in almost all pipes; some pipes with flow direction reversals do not have this maximum cross-correlation. It is important to remember that the incoming flow pattern that was used as a reference for the cross-correlation is not equal to the common flow pattern that would be applied in this model, i.e. the flow pattern of the pumping station. The cross-correlation between the pumping station flow pattern and the incoming flow pattern is less than 1. The auto-correlation of different time lags is not shown; it gives very similar results to the cross-correlation.

Figure 8.5 shows, per pipe, the percentage of the day that the flow is stagnant in the BU model. Figure 8.6 and Figure 8.7 show in a similar way the percentage of the day that the flow is laminar ($Re < 2000$) and turbulent ($Re > 4000$), respectively. Especially the connection lines and the ends of branches show stagnancy for a significant part of the day. The looped part of the network where flow direction reversals occur (Figure 8.8) has mainly laminar flows; 45% of the distribution pipes experience laminar flow for 80% of the time or more. The larger pipes that are the main feeding lines show primarily turbulent flow; 6% of the pipes experience turbulent flow for at least 80% of the time. The TD model has no stagnant flows; instead, most of the network experiences laminar flows, including the connection lines. The representation of turbulent flows in the TD model resembles that of the BU model.

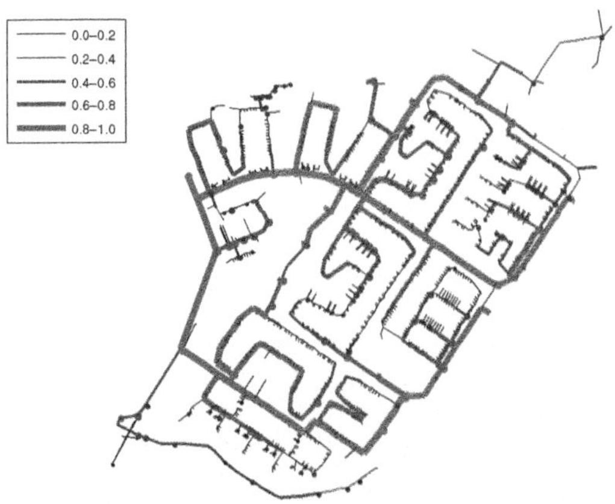

Figure 8.4 Cross-correlation of incoming flow and flow in all pipes (0.01 h time step)

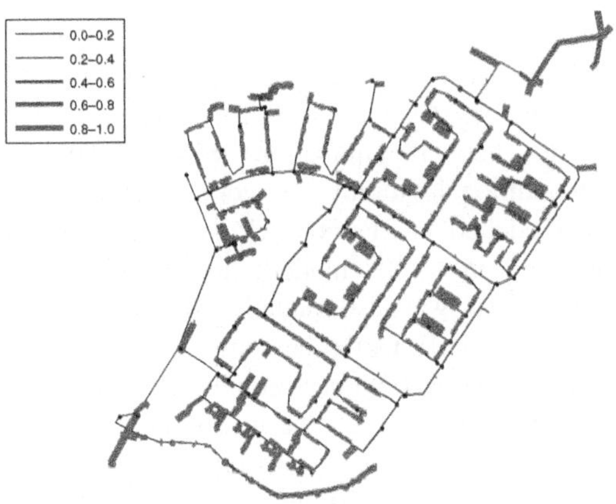

Figure 8.5 Percentage of time that pipe flow is stagnant pipes (0.01 h time step)

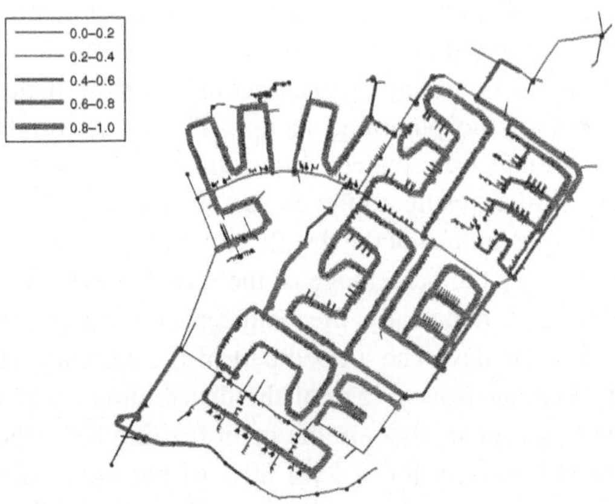

Figure 8.6 Percentage of time that pipe flow is laminar pipes (0.01 h time step)

Figure 8.7 Percentage of time that pipe flow is turbulent pipes (0.01 h time step)

Figure 8.8 Percentage of time that flow direction reverses. 50% means that 50% of the time flow is in one direction, and 50% of the time in the other direction pipes (0.01 h time step)

Figure 8.8 shows the flow direction reversals. The main feeding lines, the connection lines and the branched lines show no flow direction reversals; 50% of the pipes do show flow direction reversals and 8% of the pipes show it for more than 25% of the time. In the TD model only a few flow direction reversals occur, that is on those locations where the BU model shows more than 25% of flow direction reversals.

Figure 8.9 shows the maximum travel time on a day (day 2 of the simulation run) from the entrance of this network. Travel time from the pumping station to the network is not taken into account. With this network, 15% of the pipes show more than 24 hours of travel time. Figure 8.11 shows the difference in maximum travel time between the TD and BU models. The BU model leads to shorter travel times in part of the network and longer travel times in other parts. This model shows more extremes (>40 h).

Figure 8.10 shows the maximum flow velocity on a day; for reasons of readability the connection lines are not demonstrated here. Seven per cent of the pipes experience a maximum velocity >0.2 m/s; 70% of the pipes have maximum velocities <0.1 m/s. Figure 8.12 shows the difference in maximum flow velocity between the TD and BU model; the velocity in the connection lines is shown here as well. The BU model shows equal or higher velocities. Especially in the branched lines (including the connection lines) that feed a limited number of homes, the BU model shows much higher velocities.

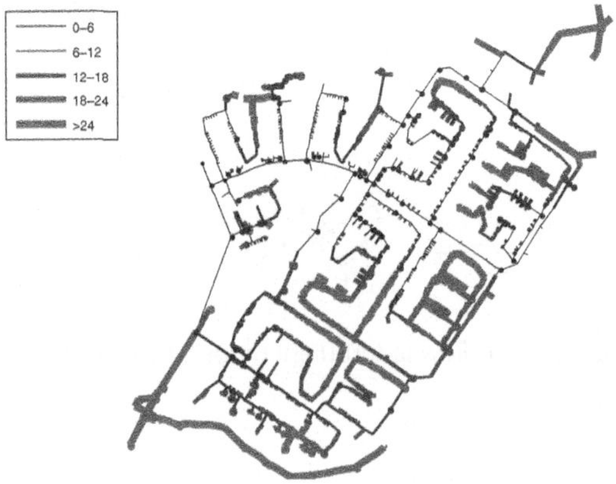

Figure 8.9 Maximum travel time (h) per pipe pipes (0.01 h time step)

Figure 8.10 Maximum flow velocity (m/s) per main pipe; connection lines are not computed pipes (0.01 h time step)

To summarise, the BU model provides insight into the extremes at the periphery of the network. It shows the longest travel times and highest maximum velocities in the branched pipes that serve a limited number of connections. Also, it shows that flow directions reversals occur often in looped systems. The figures show that in the main feeding lines a TD model can be applied, but in the periphery a BU model gives different results. The BU approach leads to realistic demand patterns, as shown by the flow measurement and tracer tests in Chapter 6 and the study on the self-cleaning velocity in Chapter 7. This is partly due to the possibility of using a smaller time step with the new approach of demand modelling, and partly due to the fact that SIMDEUM provides demand patterns with lower auto-correlation in the periphery of the network.

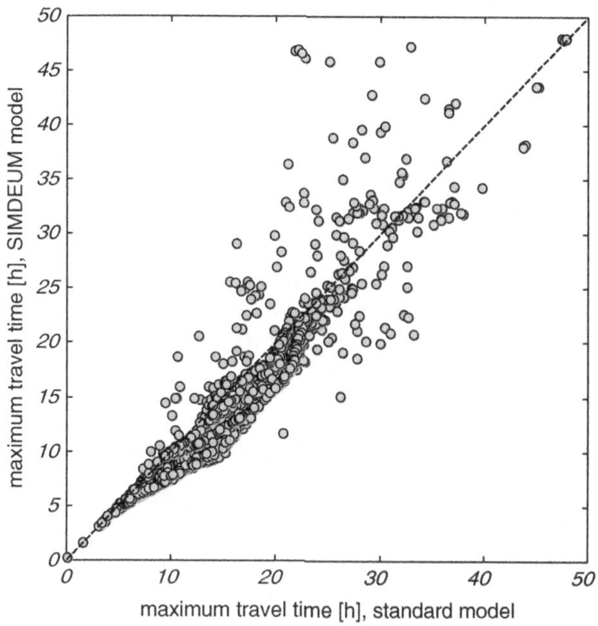

Figure 8.11 Maximum travel time (h) in main pipes in standard (TD) model and SIMDEUM (BU) model

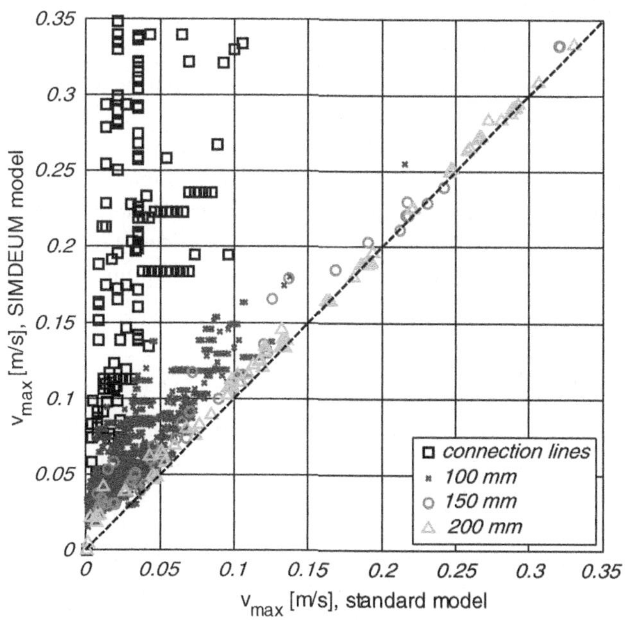

Figure 8.12 Maximum flow velocity (m/s) in all pipes in standard (TD) model and SIMDEUM (BU) mode

8.6 APPLICATIONS OF SIMDEUM IN RESEARCH ON WATER QUALITY IN THE DWDS

8.6.1 Fouling prediction tool

There is interest in a DWSD fouling prediction tool, i.e. a tool that can predict what part of the network will foul the most and needs the most cleaning. The Water Quality Distribution Modelling Toolbox (WQDMTB v4.3; Ryan et al. 2008) was tested (Vogelaar and Blokker 2010). This approach has the potential of becoming a proper fouling prediction tool. It showed that particles characteristics are still not fully understood. Moreover, it should be possible to classify the fouling rate of a DWDS based on its hydraulic conditions. Particle accumulation is mainly related to average flow velocities; the mobilisation of particles is related to maximum flow velocities. SIMDEUM demands have an added value in the modelling of maximum flow velocities.

8.6.2 Design principles for self-cleaning networks

Design principles for self-cleaning networks are available. Using the $q\sqrt{n}$-method with 15 tap units and a design velocity of 0.4 m/s to determine the desired pipe diameter leads to self-cleaning networks. As an alternative, the maximum diurnal demands from SIMDEUM together with the self-cleaning velocity of 0.25 m/s can be used to determine the maximum pipe diameters in a self-cleaning network (Chapter 7). With SIMDEUM, the maximum flow that occurs once per 10 years can be established and then this can be used to determine the minimum pipe diameter that will ensure a certain maximum pressure at all times. These design principles apply to the periphery of the DWDS, where there is a lot of stagnant water during the day.

8.6.3 Maximum travel time

There are no guidelines for the maximum travel time as it is not yet clear how exactly the water quality deteriorates over time. The deterioration depends on the initial water quality, pipe material and temperature. The temperature in the DWDS will approach the soil temperature; how fast the soil temperature is reached depends on the pipe material and the flow velocity. In practice, however, the travel times are long enough for the drinking water temperature to reach the soil temperature around the distribution mains (Blokker and Pieterse-Quirijns, in prep.). Thus, for temperature modelling SIMDEUM is not required; the conventional way of demand allocation suffices. For determining maximum travel times, SIMDEUM is recommended because it demonstrates the extremes better than an approach with non-stochastic demands.

8.6.4 Sensor placement

On the matter of water quality control, sensor placement is an important issue. Water quality sensors are being developed and installed. For a good hydraulic model, it is very important to decide where to place the sensors and to be able to interpret the measurements (*Optimal sensor location and source location inversion*). The required reliability of a water demand model in finding the optimal source location, and in-source location inversion has been investigated to a limited extent. It may not be easy to determine if model uncertainties are related to uncertainty in demand or other unknowns, due to the limited number of sensors in a network. In case people change their water consumption when a contamination is at hand, the standard demand modelling will no longer be reliable. People may still flush the toilet, but stop using the water for consumption or even for showering. A scenario study with SIMDEUM demands may be required.

8.7 PRACTICAL APPLICATIONS OF SIMDEUM

8.7.1 Water demand management

Water demand management is concerned with strategies for the reduction of water demand. Several approaches are possible to determine if a measure (e.g. a price increase or installing a water meter), has an effect on water demand. One way is to measure total water demand before and after the measure takes place and the difference must have been the effect of the measure that was taken. Another way is to ask people how the measure affects their behaviour: do they take shorter showers or have they installed a water-saving appliance? The advantage of the latter is that the results of the study are more transferable. The measure of price increase may have a less meaningful result in a region where people already installed water-saving appliances in the past. The SIMDEUM approach of applying insight into end uses can have added value in water demand management.

8.7.2 Scenario studies

The SIMDEUM model is based on independent statistical information about in-home installations and residents. Therefore, the influence of an ageing population on the total and peak water demand, of the decrease in household size, or of the replacement of old appliances with new ones can easily be determined. The end-use model thus enables quantitative scenario studies. The following scenario studies are foreseen.

In the future, water consumption may change. The change may be technology driven, e.g. new sanitation concepts like waterless toilets could emerge; it may be driven by demographics, e.g. an ageing population or a population drift from rural into the urban areas; or, it may be driven by the changing climate or other factors. The effects of these different drivers on people's behaviour with respect to water use need to be studied. SIMDEUM can be used to quantify the effects on changing water use on total and peak water demands.

Another possible scenario is that a DWDS will be supplying less drinking water due to specialisation of networks. For example, a separate drinking water network and grey water network (for flushing toilet, watering the garden, washing clothes) are installed. Or, a separate cold water network and hot water network (for showering and potentially for hot fill washing machines and dishwashers) are installed. By using SIMDEUM plus a network solver, the effect on hydraulics and water quality in the DWDS can be studied.

8.7.3 Risk analysis of contamination through ingestion

To verify the water quality at the customer's tap, a sampling programme is used. In the Netherlands, guidelines (VROM 2005) state that drinking water samples for metal analysis should be taken randomly over the network and randomly over the (work) day. In practice, this randomness is not easy to reach (Slaats et al. 2007). Also, because people must be home for the inspection, most often water is already used before the sample is taken and thus the water that remained overnight in the pipes is never being sampled. A sampling programme based on actual water use may be more appropriate (Davis and Janke 2009).

Water demand patterns per end use can be used to study the exposure of humans to water-borne contaminants at the building fixture level, based on the movement of contaminated water from the street main and into and through the building (Boccelli et al. 2006; Grayman and Buchberger 2006); SIMDEUM provides water demand at the fixture level.

8.7.4 Dimensioning plumbing and water heaters

Since SIMDEUM describes the water demand at the fixture level, the model can also be used to simulate water demand within a home. Knowledge of maximum demand supports the pipe diameter selection. Knowledge of total domestic hot water demand, in different timeframes, supports the choice of water heater. The Dutch installation organisations Uneto-VNI and TVVL have supported the development of SIMDEUM and have used this model to determine the design parameters for different types of homes and apartment buildings (Blokker and van der Schee 2006; Blokker et al. 2007; Pieterse-Quirijns 2008) as well as commercial buildings (Pieterse-Quirijns 2010).

8.7.5 Water-related energy use

SIMDEUM can provide insight into the energy that is used to heat water. SIMDEUM can also provide insight into the energy capacity of the heated drinking water that is leaving the house through the drain. It is, therefore, one of the building blocks in the research into water-related energy use.

The main reason that washing machines in Europe have decreased their water use is that an Energy Label was enforced. The less water that needs to be heated, the less energy is used and the easier it is to get an A-label. In parts of the world where washing machines do not heat the water, the water use is much higher per washing cycle. The recommendation that people take shorter showers is also more related to the energy used to heat the water and the cost related to water heating, than to the water use itself.

Now that newly built homes become more and more energy efficient due to building insulation, the heating of drinking water is a substantial part of the total energy use in Dutch households. It is a waste of energy to flush the heat down the drain. Heat exchangers to recover the heat from the shower water are on the market (http://www.shower-save.com/). They are tested with standard water demands. SIMDEUM demand patterns can provide insight into the heat recovery in realistic residential water use scenarios.

8.7.6 Practical application of SIMDEUM in network modelling

Most Dutch drinking water companies are upgrading their transport network models to an all-pipes network model. With more detail in pipes and demand nodes, and a wish for more insight into water quality in the DWDS, the Dutch water companies have shown an interest in investing in a demand pattern generator which provides water demand patterns with more temporal and spatial variation than the currently used demand patterns that are based on flow measurements at pumping stations.

The study of a bottom-up approach of demand allocation in the small Benthuizen test area (Chapter 6) showed that this approach leads to a good understanding of average and variance of residence times. This comes at a cost. Compared to the conventional top-down approach of demand allocation, the new model approach leads to larger hydraulic models with more nodes, more pipes and more numerous demand patterns. In the test model, an extra node, pipe and demand pattern were added for each individual home. Using a smaller time step means that simulations take longer; a time step of 1 minute instead of 1 hour, approximately leads to a 60 times longer simulation. In the test area, simulations were still very rapid. To determine the variance of the residence time, multiple simulation runs are required. This also means a longer total simulation time. With the still increasing computer capacity, the problem of more demanding hydraulic models will diminish. Another practical issue is that the analysis of the set of results cannot be done in a hydraulic network solver, but needs to be done elsewhere.

The sensitivity analysis of Chapter 6 showed that for the specific purpose of determining residence times for the selected locations, there is no need to run the hydraulic model at as small a time step as one second; a 15 minute time step suffices. Also, some spatial aggregation is permitted; it is not required that each home has its own demand node in the hydraulic model. A stochastic modelling approach of 10 different simulations is enough to get a good feel for the 95% confidence interval. These findings can help to limit the model increase.

The construction of an "all pipes" network model of the larger (Zandvoort) network in Chapter 6 meant an effort. This effort was not specifically done for the study. Most Dutch water companies are migrating to using all pipes network models as they can automatically be generated from the GIS systems. Filling the model with the appropriate water demand patterns and running the simulations was automated and took little effort. The study in this more practical network showed that a bottom-up approach of demand allocation in real networks is feasible without the need for calibration on water demands.

SIMDEUM demand patterns have successfully been applied in EPANET (Chapter 6) and in the network solvers that are used by the Dutch drinking water companies: ALEID (Blokker et al. 2006; Blokker and Vogelaar 2007; Blokker and Beverloo 2008), InfoWorks® (Chapter 6) and SynerGEE® (Blokker et al. 2009a). The commercially available network solvers do have some practical limitations: InfoWorks (version 10.5.8) does not accept pattern time steps of less than 1 min, and SynerGEE (4.0.3) does not have a fast demand pattern importing method.

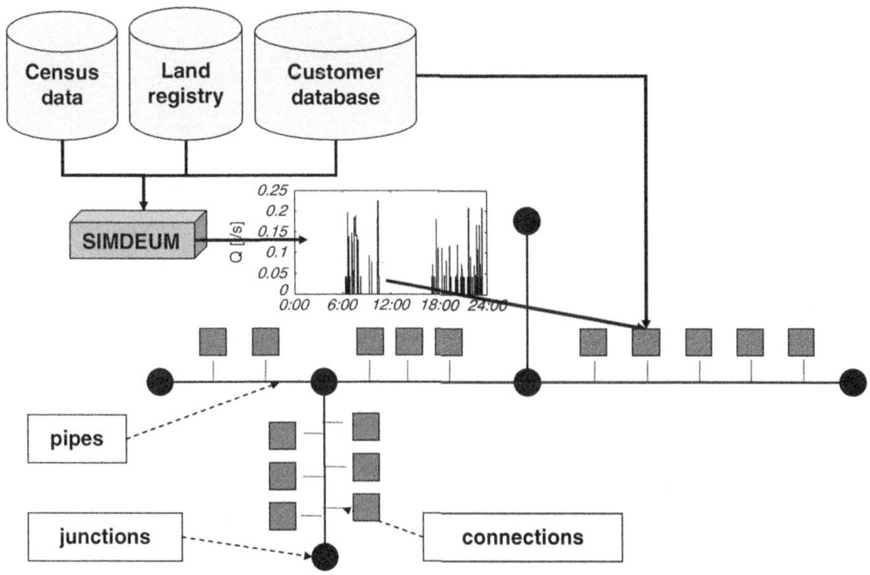

Figure 8.13 The future of network model building: applying SIMDEUM demand patterns to an all-pipes network model, using external databases

The envisioned future of hydraulic network modelling (Blokker et al. 2009a) is that by connecting the water company's customer database (with information on annual water demand), census data (household size and age of people in a neighbourhood), and the land registry database (is a property residential or a business, e.g. a hair dresser, office, etc.), a SIMDEUM based pattern generator can provide demand patterns which can be linked to the network model, as schematically visualised in Figure 8.13. Depending on the application (checking on pressures, determining travel times, processing sensor data, etc.) the model may need to be properly skeletonised to limit calculation time. The skeletonisation process would have to include demand re-allocation. Saldarriaga et al. (2009) have shown that demand model skeletonisation is allowed with respect to pressures in the pipes, but it has a manifest effect on the water quality model results. This means that limited skeletonisation is allowed in water quality modelling.

The Dutch water companies Dunea, PWN and Waternet are interested in calculating their networks with the appropriate demand patterns for work days, weekends, winter and summer days, and out of the ordinary situations. That means that not only the average flows and the maximum day flows are studied, but, for example, also the situation that people may change their water use in the event of a contamination. Also, the water companies would like to use specific demand patterns for residential and commercial areas.

As a first step, a library of water demand patterns for different categories of water demand and for different situations can be created. The current Dutch studies of residential water use (between 1992 and 2007 (Foekema et al. 2008) only show a minor change in average water use. Also, the time-budget surveys of 1995 and 2005 (SCP 1995; SCP 2005) show minor differences in sleep-wake rhythms. This means that on average there are no major changes and the input data into SIMDEUM need not be determined very often. Larger variations can be expected from the day of the week or from seasonal effects. Water demand patterns on a maximum day can be constructed with the knowledge of these external influences. SIMDEUM can provide residential and non-residential water demands. Within the non-residential category, building types should be defined that are distinguished by their diurnal water demand pattern.

Chapter 9

Summary and conclusions

9.1 INTRODUCTION

The objectives of this book were, firstly, to develop and validate the water demand model SIMDEUM; secondly, to apply SIMDEUM in case studies; and, thirdly, to evaluate the added value of SIMDEUM in hydraulic and water quality network modelling. The accomplishment of these objectives is discussed in this chapter.

9.2 DEVELOPING AND VALIDATING SIMDEUM: A SIMULATION MODEL FOR DRINKING WATER DEMAND

The model SIMDEUM was developed. SIMDEUM is an acronym for "SIMulation of Demand; an End-Use Model". It simulates water demand over the day with a one second time step and per fixture. The model is based on information of the end use at the fixture level. This means information of the fixtures: what is the (maximum) flow of water use, what is the duration of water use? It also means information of the end users: how long do people open a tap, when during the day do they open a tap, and how many people have access to the tap? The sources of information are reliable independent studies that, moreover, are frequently repeated. The first source is the study of residential water use, which has been commissioned by Vewin every three years since 1992. The second source is the time budget survey of the SCP which has been repeated every five years since 1975 and which can be freely downloaded of the Internet. The third source is the public information from the Statistics Bureau CBS on households (size and age distribution) per neighbourhood. Next to that, some technical information from manufacturers of water-using appliances was used.

SIMDEUM is a stochastic model. This means that the model generates diurnal water demand patterns from different statistical probability distributions of the time of a water-use event, the duration of water use and the flow. Every simulated pattern, therefore, is different. To validate and interpret the model results, statistical parameters are used. For example, the extremes of maximum demand are evaluated. Chapter 3 describes the basics of the model and the input data.

SIMDEUM was validated with measurements of individual homes in Amsterdam (Chapter 3), a street of 140 homes in Benthuizen (Chapter 6) and a neighbourhood of 1000 homes in Zandvoort (Chapter 6). SIMDEUM predicts the water demand well with respect to average and peak demands and the diurnal pattern.

Since SIMDEUM was based on information of the underlying principle of water use, a variety of applications emerges. SIMDEUM was developed to simulate Dutch residential water demand, but it can be applied for (Dutch) offices, hotels and nursing homes as well (Chapter 4). SIMDEUM can also simulate residential USA water demand, as was shown in Chapter 5 for 21 homes in Milford, Ohio.

9.3 APPLYING SIMDEUM IN CASE STUDIES: THE STUDY OF RESIDENCE TIMES AND PARTICLES IN THE DISTRIBUTION NETWORK

The demand patterns from SIMDEUM were applied in network models in several case studies, three of which meant further validation of SIMDEUM with field measurements. Two case studies were related to tracer tests in two distribution networks (Chapter 6). In these studies, SIMDEUM was used to predict residence times in a distribution network. The third case study was related to the fouling of a distribution network (Chapter 7). The results of turbidity measurements during flushing were paired with the resulting maximum flow velocities during normal operation in a detailed hydraulic network model with SIMDEUM demand patterns. This showed the effect of daily maximum velocity on the local fouling of a water main.

SIMDEUM demand patterns were applied in several hydraulic network solvers that are used in research and by the Dutch drinking water companies, viz. EPANET, ALEID, InfoWorks® and SynerGEE®.

The application of SIMDEUM in hydraulic modelling means a different approach of demand allocation: bottom-up instead of top-down. Top-down means that a demand multiplier pattern (for example, based on measured pumping station flows) is applied to the demand nodes, and the corresponding base demands are related to annual water use. This implies a strong cross-correlation of demands between nodes. Bottom-up means that for each individual home a unique water demand pattern is generated, and that an aggregated pattern is allocated to the demand nodes. Also, the application of SIMDEUM in hydraulic modelling means that stochastic modelling and interpretation are required. Different sets of random demand patterns lead to differences in total demand, in residence times and maximum velocities.

9.4 EVALUATING ADDED VALUE OF SIMDEUM: COMPARING SIMDEUM TO EXISTING MODELS

In Chapter 6, SIMDEUM-based models are being compared to the conventional way of hydraulic modelling, i.e. using a demand pattern based on flow measurements of a supply area. The comparison was done on the total flow of the day of a distribution network and on calculated residence times in the network. Both the conventional and the new SIMDEUM-based model were compared to measured flow patterns and measured residence times in a tracer study.

Compared to the conventional model, the SIMDEUM-based model resulted in the total flow and residence times that were at least as accurate. SIMDEUM leads to similar average values, but a larger variation than the conventional model. This larger variation resembles the actual (measured) variation and thus, hydraulic models with SIMDEUM demand patterns yield extra information. The application of SIMDEUM has an added value in residence time studies because for various water quality processes in the network, the maximum residence time is more important than the average.

Next to residence time, solute transport was also considered, in particular the shape of the EC (electrical conductivity) pulse of the added tracer. At one of the measurement locations, the EC pulse was clearly split into several pulses. This was caused by the fact that the water was supplied from two sides. With the SIMDEUM-based model such a splitting of pulses did occur, whereas in the conventional model it never did.

In Chapter 5, the SIMDEUM approach was compared to the approach of the existing PRP in the case studie of Milford. Both models simulate water demand on a spatial scale of the individual home and a temporal scale of as low as one second. Both models are based on the fact that single water use events can be described as block pulses (with a certain duration and flow and occurring at a certain moment in the day), which can be summed to generate the flow in a home. The PRP model describes the water use event time, duration and flow with three probability distributions that are found from flow measurements. SIMDEUM describes the water use event time, duration and flow with many end-use dependent probability distributions that are found from the analysis of surveys. The input parameters for SIMDEUM for Milford could easily be found in the literature.

The results of both models compare well with the measurement data. An issue is that it is difficult to determine how well the PRP model performs compared to the measurements, since the simulation parameters were derived from the same measurements.

The two models each have their own strengths and weaknesses and thus their own application areas. The PRP model is intended to be a descriptive model and, with a limited set of input parameters, it can give analytical solutions to water demand related issues. SIMDEUM is a predictive model. It enables the explanation of the diurnal pattern and how it may alter with changing household occupancy, water-using appliances or habits with respect to water use, for example due to climate change. SIMDEUM can be used in the design phase of in-home plumbing and the distribution network, and with scenario studies. SIMDEUM provides insight in the water use, including hot water use, of the various fixtures.

9.5 EVALUATING ADDED VALUE OF SIMDEUM: NEW WAY OF DEMAND MODELLING IS REQUIRED FOR MODELLING WATER QUALITY IN THE DISTRIBUTION NETWORK

Chapter 2 points out that the application of a new type of water demand model is required in the study and modelling of water quality changes in the distribution network, especially in the periphery of the distribution network. This part of the network is characterised by long residence times, low flow velocities and water demand patterns that show limited cross-correlation between demand nodes. Water quality is mainly important in this part of the network since this part of the network serves customers and the water has the longest retention time within the distribution network.

The most relevant water quality parameters are the risk of discoloured water, the dispersal of a contaminant over the distribution network, chlorine decay and an increase in disinfection by-products, the growth of the biofilm and microbiological organisms. The discolouration risk is related to the number of particles coming from the pumping station that accumulate in the network and can be mobilised and then lead to brown water complaints. The hydraulics influence the accumulation and mobilisation processes: accumulation occurs during low flow velocities, mobilisation takes place at high flow velocities. The chemical and biological processes in the distribution network are related to pipe material properties, the water quality from the pumping station and to physical processes in the networks such as water temperature, contact time (or residence time), flow velocity, duration of stagnant flows (which allows for more interaction with the wall than during low flows), etcetera. The contaminant dispersal is influenced by a combination of the water demands at all time steps and all demand nodes in the network. Therefore, water quality processes in water distribution networks are strongly influenced by the hydraulics.

The selection of a demand model (including the scale of temporal and spatial aggregation) influences the important water quality modelling parameters cross- and auto-correlation, the probability of stagnant water, laminar and turbulent flows, and the value of the maximum flow velocity. Chapter 8 shows, for a typical (Dutch) distribution network, i.e. a residential area with many terraced houses and a looped distribution network, the effect of a detailed hydraulic model on these parameters. In the periphery of the distribution network, the difference between the new way of demand allocation (bottom-up, BU) and the conventional way of demand allocation (top-down, TD) is apparent: the cross- and auto-correlation are much lower than the TD model assumes, in the looped parts the flow is not uni-directional but constantly changes direction, stagnancy actually occurs often and turbulent flows occur seldom. Every now and then a high flow velocity occurs, which is underestimated with the TD model. To be able to determine if a potential contamination has been removed from the entire distribution system or if chlorine levels are sufficient at the end of the network, a model is required with water demand patterns with a high (spatial and temporal) level of detail. Also, the prediction of particle accumulation in the network requires such a detailed water demand model. Chapter 7 has shown that a bottom-up approach of demand allocation has an added value in the research of the relationship between particle processes and hydraulics.

The purpose of this book is to get a better understanding of the hydraulics in water distribution networks by supplying a solid water demand model. The objective is to model the flow velocities in the pipe system at any moment in the day in such a way that instantaneous velocities (in particular the maximum velocity) can be related to particle processes, that the residence time and temperature can be related to chemical and biological processes in the network and the contaminant dispersal in the network can be predicted well. An improved hydraulics model will support the application and development of water quality models.

References

Ackers, J., Brandt, M., and Powell, J. (2001). *Hydraulic Characterisation of Deposits and Review of Sediment Modelling*, UK Water Industry Research, London, Uk. 01/DW/03/18.

Alcocer-Yamanaka, V., and Tzatchkov, V. (2009). "Spectral analysis of instantaneous residential water demand time series." *CCWI 2009 'Integrating water systems'*, Sheffield, 503–508.

Alcocer-Yamanaka, V. H., Tzatchkov, V. G., and Buchberger, S. G. (2006). "Instantaneous water demand parameter estimation from coarse meter readings." Water Distribution System Analysis #8., ASCE, Cincinnati, Ohio, USA, 51–51.

Alvisi, S., Franchini, M., and Marinelli, A. (2003). "A Stochastic Model for Representing Drinking Water Demand at Residential Level." *Water Resources Management*, 17(3), 197–222.

Austin, R. G., Waanders, B. v. B., McKenna, S., and Choi, C. Y. (2008). "Mixing at cross junctions in water distribution systems. II: experimental study." *Journal of Water Resources Planning and Management*, 134(3), 295–302.

Babayan, A. V., Savic, D. A., and Walters, G. A. (2005). "Multiobjective optimization for the least-cost design of water distribution system under correlated uncertain parameters." Impacts of Global Climate Change; 2005 World water and environmental resources congress, R. Walton, ed., American Society of Civil Engineers, Anchorage, Alaska.

Baggelaar, P. K., and Geudens, P. J. J. G. (2008). *Prognoses landelijke drinkwatervraag tot 2025* (Prognosis national water demand until 2025), Vewin. 2008/85/6222.

Berry, J. W., Hart, W. E., Philips, C. A., Uber, J. G., and Walski, T. M. (2005). "Water Quality Sensor Placement in Water Networks with Budget Constraints." Impacts of Global Climate Change; 2005 World water and environmental resources congress, R. Walton, ed., American Society of Civil Engineers, Anchorage, Alaska, 60.

Beuken, R. H. S., Lavooij, C. S. W., Bosch, A., and Schaap, P. G. (2006). "Low leakage in the Netherlands confirmed." Water Distribution System Analysis #8, American Society of Civil Engineers, Cincinnati, Ohio, USA.

Blokker, E. J. M. (2006). *Modelleren van afnamepatronen; beschrijving en evaluatie van simulatiemodel SIMDEUM* (Modelling of demand patterns; description and evaluation of simulation model SIMDEUM), Kiwa N.V., Nieuwegein. BTO 2006.010.

Blokker, E. J. M., and van der Schee, W. (2006). "Simulation of water demands provides insight." CIB W062 Symposium 2006 - water supply and drainage for buildings, Taipei, Taiwan.

Blokker, E. J. M., Vreeburg, J. H. G., and Vogelaar, A. J. (2006). "Combining the probabilistic demand model SIMDEUM with a network model." Water Distribution System Analysis #8, American Society of Civil Engineers, Cincinnati, Ohio, USA.

Blokker, E. J. M., Doldersum, R., Lansbergen, A., van der Schee, W., and Scheffer, W. (2007). *Rekenregels voor dimensionering van leidingwaterinstallaties* (Design rules for dimensioning drinking water installations), Kiwa Water Research, Nieuwegein. KWR 06.104.

Blokker, E. J. M., and Vogelaar, A. J. (2007). *Toepassing van SIMDEUM in het leidingnetmodel van Franeker* (Application of SIMDEUM in the network model of Franeker), Kiwa Water Research B.V., Nieuwegein. BTO 2007.006 (s).

Blokker, E. J. M., and Beverloo, H. (2008). *Verandering van waterkwaliteit in het distributienet; Metingen in Rosmalen in vemaasd en vertakt netten* (Changes of water quality in the distribution network; measurements in Rosmalen in branched and looped systems), KWR, Nieuwegein. BTO 2008.039.

Blokker, E. J. M., Buchberger, S. G., Vreeburg, J. H. G., and van Dijk, J. C. (2008). "Comparison of water demand models: PRP and SIMDEUM applied to Milford, Ohio, data." WDSA 2008, J. E. van Zyl, A. A. Ilemobade, and H. E. Jacobs, eds., Kruger National Park, South Africa, 182–195.

Blokker, E. J. M., Groot, R. d., Arfman, M. K., and Roer, M. v. d. (2009a). *Toekomst van leidingnetmodelleren* (The future of hydraulic network modelling), KWR, Nieuwegein. KWR 09.090 (DPW).

Blokker, E. J. M., Schaap, P. G., and Vreeburg, J. H. G. (2009b). "Self-cleaning networks put to the test." Chapter, Strategic Asset Management of Water Supply and Wastewater Infrastructures; Invited papers from the IWA Leading edge conference on strategic asset management (LESAM), Lisbon, October 2007, H. Alegre and M. d. C. Almeida, eds., 407–417.

Blokker, E. J. M., and Pieterse-Quirijns, E. J. (in prep.). "Modelling temperature in the Drinking Water Distribution System."

Boccelli, D. L., Moll, D., Uber, J. G., Blount, B., and Cardinali, F. (2006). "Characterization of exposure to trihalomethanes in a distribution system using modelling and epidemiologic methods: preliminary analysis and descriptive statistics." Water Distribution System Analysis #8, American Society of Civil Engineers, Cincinnati, Ohio, USA.

Bosch, N., Deelen, A., and Euwals, R. "Is Part-Time Employment Here to Stay? Evidence from the Dutch Labour Force Survey 1992–2005." *SSRN eLibrary*, doi:10.1111/j.0042-7092.2007.00700.x.

Bowden, G. J., Nixon, J. B., Dandy, G. C., Maier, H. R., and Holmes, M. (2006). "Forecasting chlorine residuals in a water distribution system using a general regression neural network." *Mathematical and Computer Modelling*, 44, 469484.

Boxall, J. B., Skipworth, P. J., and Saul, A. J. (2001). "A novel approach to modelling sediment movement in distribution mains based on particle characteristics." the Computing and Control in the Water Industry Conference, A.A. Balkema Publishers, De Montfort University, UK.

Boxall, J. B., and Saul, A. J. (2005). "Modeling Discoloration in Potable Water Distribution Systems." *Journal of Environmental Engineering*, 131(5), 716–725.

Buchberger, S. G., and Wu, L. (1995). "Model for Instantaneous Residential Water Demands." *Journal of Hydraulic Engineering*, 121(3), 232–246.

Buchberger, S. G., and Wells, G. J. (1996). "Intensity, duration and frequency of residential water demands." *Journal of Water Resources Planning and Management*, 122(1), 11–19.

Buchberger, S. G., and Lee, Y. (1999). "Evidence supporting the Poisson pulse hypothesis for residential water demands." *CCWI: International Conference on Computing and Control for the Water Industry*, Exeter, United Kingdom.

Buchberger, S. G., Carter, J. T., Lee, Y. H., and Schade, T. G. (2003). *Random demands, travel times, and water quality in dead ends*, AWWARF, Denver, Colorado. AWWARF Report 90963F.

Buchberger, S. G., Blokker, E. J. M., and Vreeburg, J. H. G. (2008). "Sizes for Self-Cleaning Pipes in Municipal Water Supply Systems." WDSA 2008, J. E. van Zyl, A. A. Ilemobade, and H. E. Jacobs, eds., ASCE, Kruger Park, South Africa, 338–347.

Buchberger, S. G., and Li, Z. (2009). "Effect of spatial correlation among water demand on contamination transport in a pipe network." *CCWI 2009 'Integrating water systems'*, Sheffield, 377–382.

CBS http://statline.cbs.nl, December 2007.

Centre for Time Use Research http://www.timeuse.org, December 2007.

Clark, R. M., Grayman, W. M., Buchberger, S. G., Hartman, D., and Lee, Y. H. (2004). "Drinking Water Distribution Systems, An Overview." Chapter 4, Handbook on Water Supply Systems: Security and Safety, L. W. Mays, ed., McGraw-Hill.

Cobacho, R., Arregui, F., Gascón, L., and Cabrera, E. (2004). "Low-flow devices in Spain: how efficient are they in fact? An accurate way of calculation." *Water Science and Technology: Water Supply*, 4(3), 91–102.

Cook, D. M., Spain, P., and Wilkes, C. (2009). "Investigating relationships between discolouration customer contact frequencies and mains properties in drinking water distribution systems." *CCWI 2009 'Integrating water systems'*, Sheffield, 391–395.

Davis, M. J., and Janke, R. (2009). "Development of a Probabilistic Timing Model for the Ingestion of Tap Water." *Journal of Water Resources Planning and Management*, 135(5), 397–405.

EPA. (2002). *Effects of water age on distribution system water quality*, prepared by AWWA with assistance from Economic and Engineering Services, Inc. EPA White paper.

Filion, Y. R., Karney, B. W., and Adams, B. J. (2005). "Stochasticity of demand and probabilistic performance of water networks." Impacts of Global Climate Change; 2005 World water and environmental resources congress, R. Walton, ed., American Society of Civil Engineers, Anchorage, Alaska, 49.

Filion, Y. R., Karney, B. W., Moughton, L. J., Buchberger, S. G., and Adams, B. J. (2006). "Cross correlation analysis of residential demand in the city of Milford, Ohio." Water Distribution System Analysis #8, American Society of Civil Engineers, Cincinnati, Ohio, USA.

Foekema, H., van Thiel, L., and Lettinga, B. (2008). *Watergebruik thuis 2007*, TNS NIPO, Amsterdam. in opdracht van VEWIN.

Foekema, H., and van Thiel, L. (2011). *Watergebruik thuis 2010* (Water use at home 2010), TNS NIPO, Amsterdam. in opdracht van VEWIN.

García, V., García-Bartual, R., Cabrera, E., Arregui, F., and García-Serra, J. (2004). "Stochastic model to evaluate residential water demands." *Journal of water resources planning and management*, 130(5), 386–394.

Geudens, P. J. J. G. (2008). *Water supply statistics 2007*, Vewin, Rijswijk. 2008/83/6259.

Gill, W. N., and Sankarasubramanian, R. (1970). "Exact analysis of unsteady convective diffusion." *Proceedings of the Royal Society of London Series A*, 316, 341–350.

Gleick, P. H., Haasz, D., Henges-Jeck, C., Srinivasan, V., Wolff, G., Kao Cushing, K., and Mann, A. (2003). *Waste not, want not: the potential for urban water conservation in California*, Pacific Institute for Studies in Development, Environment, and Security, Oakland, California, USA.

Grayman, W. M., and Buchberger, S. G. (2006). "Fixture-level human exposure calculation model." Water Distribution System Analysis #8, American Society of Civil Engineers, Cincinnati, Ohio, USA.

Grayman, W. M., Speight, V. L., and Uber, J. G. (2006). "Using Monte-Carlo simulation to evaluate alternative water quality sampling plans." Water Distribution System Analysis #8, American Society of Civil Engineers, Cincinnati, Ohio, USA.

Guercio, R., Magini, R., and Pallavicini, I. (2001). "Instantaneous residential water demand as stochastic point process." Chapter, Water resources management, WIT Press, Southampton, UK, 129–138.

Ho, C. K., Choi, C. Y., and McKenna, S. A. (2007). "Evaluation of Complete and Incomplete Mixing Models in Water Distribution Pipe Network Simulations." World environmental and water resources congress 2007 - Restoring our natural habitat, K. C. Kabbes, ed., ASCE, Tampa, Fl, USA.

Husband, P. S., Boxall, J. B., and Saul, A. J. (2008). "Laboratory studies investigating the processes leading to discolouration in water distribution networks." *Water Research*, 42(16), 4309–4318.

Jacobs, H. E., and Haarhoff, J. (2004). "Structure and data requirements of an end-use model for residential water demand and return flow." *Water SA*, 30(3), 293–304.

Jonkergouw, P. M. R., Khu, S.-T., Kapelan, Z. S., and Savic, D. A. (2008). "Water quality model calibration under unknown demands." *Journal of Water Resources Planning and Management*, 134(4), 326–336.

Jung, B., Boulos, P. F., and Wood, D. J. (2007). "Impacts of skeletonization on distribution system hydraulic transient models." World environmental and water resources congress 2007 - Restoring our natural habitat, K. C. Kabbes, ed., ASCE, Tampa, Fl, USA.

Kanne, P. (2005a). *Watergebruik thuis 2004, deelrapportage Waterleidingbedrijf Amsterdam* (Water use at home 2004; partial report for Water Company Amsterdam), TNS NIPO, Amsterdam. in opdracht van het waterleidingbedrijf Amsterdam.

Kanne, P. (2005b). *Watergebruik thuis 2004* (Water use at home 2004), TNS NIPO, Amsterdam. in opdracht van VEWIN.

Kapelan, Z. (2002). *Calibration of water distribution system hydraulic models*, University of Exeter. PhD thesis, p. 334.

Karney, B. W., Jung, B., and Alkozai, A. (2006). "Assessing the degree of unsteadiness in flow modeling; from physics to numerical solution." Water Distribution System Analysis #8, American Society of Civil Engineers, Cincinnati, Ohio, USA.

Lee, Y. (2004). *Mass dispersion in intermittent laminar flow*, University of Cincinnati, Cincinnati, Ohio.

Li, Z., and Buchberger, S. G. (2003). *PRP Simulator users manual*, university of Cincinnati, Cincinnati, Ohio.

Li, Z., and Buchberger, S. G. (2004). "Effect of time scale on PRP random flows in pipe network." Critical Transitions In Water And Environmental Resources Management, e. Gerald Sehlke, Donald F. Hayes, (editor), and David K. Stevens (editor), ed., ASCE, Salt Lake City, Utah.

Li, Z., Buchberger, S. G., and Tzatchkov, V. G. (2005). "Importance of disperison in network water quality modeling." Impacts of Global Climate Change; 2005 World water and environmental resources congress, R. Walton, ed., American Society of Civil Engineers, Anchorage, Alaska.

Li, Z. (2006). *Network Water Quality Modeling with Stochastic Water Demands and Mass Dispersion*, university of Cincinnati. PhD thesis, p. 165.

Li, Z., Buchberger, S. G., and Tzatchkov, V. G. (2006). "Integrating distribution network models with stochastic water demands and mass dispersion." Water Distribution System Analysis #8, American Society of Civil Engineers, Cincinnati, Ohio, USA.

Li, Z., and Buchberger, S. G. (2007). "Effects of spatial-temporal aggregation on properties of PRP random water demands." World environmental and water resources congress 2007 - Restoring our natural habitat, K. C. Kabbes, ed., ASCE, Tampa, Fl, USA.

Machell, J., Boxall, J., Saul, A., and Bramley, D. (2009). "Improved Representation of Water Age in Distribution Networks to Inform Water Quality." *Journal of Water Resources Planning and Management*, 135(5), 382–391.

Maddaus, W. O. (1987). *Water Conservation*, American Water Works Association, Denver, Colorado.

Mayer, P. W., DeOreo, W. B., Opitz, E. M., Kiefer, J. C., Davis, W. Y., Dziegielewski, B., and Nelson, J. O. (1999). *Residential End Uses of Water*, AWWARF, Denver, CO.

McInnis, D., and Karney, B. W. (1995). "Transients in distribution networks: Field tests and demand models." *Journal of Hydraulic Engineering*, 121(3), 218–231.

McKenna, S. A., Buchberger, S. G., and Tidwell, V. C. (2003). "Examining the effect of variability in short time scale demands on solute transport." World Water and Environmental Resources Congress and Related Symposia, P. D.-E. Paul Bizier, ed., Philadelphia, Pennsylvania, USA.

McKenna, S. A., van Bloemen Waanders, B., Laird, C. D., Buchberger, S. G., Li, Z., and Janke, R. (2005). "Source location inversion and the effect of stochastically varying demand." Impacts of Global Climate Change; 2005 World water and environmental resources congress, R. Walton, ed., American Society of Civil Engineers, Anchorage, Alaska, 47.

Memon, F. A., Ton-That, L., and Butler, D. (2007). "An investigation of domestic water consumption through taps and its impact on urban water flows." *Water Science and Technology: Water Supply*, 7(5–6), 69–76, doi:10.2166/ws.2007.100

Menaia, J., Coelho, S. T., Lopes, A., Fonte, E., and Palma, J. (2003). "Dependency of bulk chlorine decay rates on flow velocity in water distribution networks." *Water Science and Technology: Water Supply*, 3(1/2), 209–214.

Mesman, G. A. M., and Meerkerk, M. A. (2010). *Evaluatie ontwerprichtlijnen voor distributienetten* (Evaluation of design criteria of distribution networks), KWR, Nieuwegein. KWR 09.073.

Moughton, L. J., Buchberger, S. G., Boccelli, D. L., Filion, Y. R., and Karney, B. W. (2006). "Effect of time step and data aggregation on cross correlation of residential demands." Water Distribution System Analysis #8, American Society of Civil Engineers, Cincinnati, Ohio, USA.

Nilsson, K. A., Buchberger, S. G., and Clark, R. M. (2005). "Simulating exposures to deliberate intrusions into water distribution systems." *Journal of Water Resources Planning and Management*, 131(3), 228–236.

Paauw, J., Roossien, B., Aries, M. B. C., and Santin, O. G. (2009). "Energy Pattern Generator; Understanding the effect of user behaviour on energy systems." First European Conference on Energy Efficiency and Behaviour Maastricht.

Pardo Picazo, M. A. (2009). "Water quality model calibration using the NaCl tracer test," M.Sc., Universidad Politénica de Valencia, Valencia, Spain.

Pasha, M. F. K., and Lansey, K. (2010). "Effect of parameter uncertainty on water quality predictions in distribution systems-case study." *Journal of Hydroinformatics*, 12(1), 1–21, doi:10.2166/hydro.2010.053.

Pieterse-Quirijns, E. J. (2008). *Rekenregels voor waterverbruik in woontorens* (Design rules for water demand in apartment buildings), KWR, Nieuwegein. KWR 08.089.

Pieterse-Quirijns, E. J., Blokker, E. J. M., and Vogelaar, A. J. (2009). *Modelleren van niet-huishoudelijk waterverbruik; waterverbruik van kantoren, hotels, zorginstellingen en veehouderij* (Modelling of non-residential water demand; water demand of offices, hotels, nursing homes and famers), KWR, Nieuwegein. BTO 2009.013.

Pieterse-Quirijns, E. J. (2010). *Rekenregels voor het waterverbruik in utiliteitsbouw; Bepalen van maximum volumestroom en warmwaterverbruik met SIMDEUM* (Design rules for water demand in utility buildings; Determining maximum flow and hot water demand with SIMDEUM), KWR, Nieuwegein. KWR 2010.072.

Powell, J., Clement, J., Brandt, M., R, C., Holt, D., Grayman, W., and LeChevallier, M. (2004). *Predictive Models for Water Quality in Distribution Systems*, AWWARF, Denver, Colorado, USA. AWWARF Report 91023F.

Propato, M., and Uber, J. (2004). "Vulnerability of Water Distribution Systems to Pathogen Intrusion: How Effective Is a Disinfectant Residual?" *ENVIRONMENTAL SCIENCE & TECHNOLOGY*, 38(13), 3713–3722.

Romero-Gomez, P., Ho, C. K., and Choi, C. Y. (2008a). "Mixing at cross junctions in water distribution systems. I: numerical study." *Journal of Water Resources Planning and Management*, 134(3), 285–294.

Romero-Gomez, P., Li, Z., Choi, C. Y., Buchberger, S. G., Lansey, K. E., and Tzatchkov, V. T. (2008b). "Axial dispersion in a pressurized pipe under various flow conditions." WDSA 2008, Kruger Park, South Africa.

Rossman, L. A., Clark, R. M., and Grayman, W. M. (1994). "Modeling chlorine residuals in drinking-water distribution systems." *Journal of Environmental Engineering*, 120(4), 803–820.

Rossman, L. A. (2000). *EPANET 2 user manual*, United States Environmental Protection Agency, Cincinnati. EPA/600/R-00/057.

Ryan, G., Mathes, P., Haylock, G., Jayaratne, A., Wu, J., Noui-Mehidi, N., Grainger, C., and Nguyen, B. V. (2008). *Particles in water distribution systems*, Cooperative Research Centre for Water Quality and Treatment, Salisbury, Australia. Research report 33.

Saldarriaga, J. G., Ochoa, S., Nieto, L., and Rodirı́guez, D. (2009). "Methodology for the skeletonization of water distribution network models with demand aggregation." *CCWI 2009 'Integrating water systems'*, Sheffield, 423–429.

Schaap, P., and Drevijn, A. (2009). *Technisch accountantsrapport leidingnet per 31-12-2008* (Technical accountancy report distribution network per 31-12-2008), NV PWN Waterleidingbedrijf Noord-Holland, Heemskerk.

SCP (1995). *Tijdsbestedingsonderzoek 1995 - TBO '95* (Time budget survey 1995) *[computer file]*, Data Archiving and Network Services - DANS, Den Haag. Persistent identifier: urn:nbn:nl:ui:13-n4c-x9o.

SCP (2005). *Tijdsbestedingsonderzoek 2005 - TBO 2005* (Time budget survey 2005) *[computer file]*, Data Archiving and Network Services - DANS, Den Haag. Persistent identifier: urn:nbn:nl:ui:13-v64-rd7.

Skipworth, P. J., Machell, J., and Saul, A. J. (2002). "Empirical travel time estimation in a distribution network." *Water & Maritime Engineering*, 154(1), 41–49.

Slaats, P. G. G., Rosenthal, L. P. M., Siegers, W. G., van den Boomen, M., Beuken, R. H. S., and Vreeburg, J. H. G. (2003). *Processes involved in the generation of discolored water*, AWWARF, Denver, Co, USA. AWWARF Report 90966F.

Slaats, P. G. G., Blokker, E. J. M., and Versteegh, J. F. M. (2007). "Sampling metals at the tap: analyses of Dutch data over the period 2004–2006 " *Cost Action 637 METEAU - Metals and related substances in drinking water*, Antalya, Turkey, 61–9.

Solley, W. B., Pierce, R. R., and Perlman, H. A. (1998). "Estimated Use of Water in the United States in 1995." *U.S. Geological Survey Circular 1200.*

Taylor, G. (1953). "Dispersion of Soluble Matter in Solvent Flowing Slowly through a Tube." *Proceedings of the Royal Society of London Series A, Mathematical and Physical Sciences*, 219(1137), 186–203.

Tricarico, C. (2005). "A rehabilitation model for water distribution systems," Università degli Studi di Cassino, Cassino.

Tzatchkov, V. G., Aldama, A. A., and Arreguin, F. I. (2002). "Advection-dispersion-reaction modeling in water distribution networks." *Journal of Water Resources Planning and Management*, 131(3), 334–342.

Tzatchkov, V. G., and Buchberger, S. G. (2006). "Stochastic demand generated unsteady flow in water distribution networks." Water Distribution System Analysis #8, American Society of Civil Engineers, Cincinnati, Ohio, USA.

U.S. Census Bureau http://www.milfordohio.org/pdf/community/Census_Statistics.pdf, April 2008.

U.S. Department of Labor; Bureau of Labor Statistics http://stats.bls.gov/tus/, April 2008.

USEPA. (2006). *Initial Distribution System Evaluation Guidance Manual for Final Stage 2 Disinfection Byproducts*, Office of Water. EPA 815-B-06-002.

Vieira, P., Almeida, M. d. C., Baptista, J. M., and Ribeiro, R. (2007). "Household water use: a Portugese field study." *Water Science and Technology: Water Supply*, 7(5–6), 193–202, doi:10.2166/ws.2007.098.

Vogelaar, A. J., and Blokker, E. J. M. (2004). *Tappatronen en Leidingdimensionering; Inventarisatie van kentallen en rekenmethoden* (Demand patterns and dimensioning of pipe sizes; Listing of parameters and methods), Kiwa N.V., Nieuwegein. BTO 2004.026.

Vogelaar, A. J., and Blokker, E. J. M. (2010). *Particle Sediment Modelling. Test and analysis of programme WQDMTB v4.3*, KWR, Nieuwegein. BTO 2010.011.

Vreeburg, J. (2010). *Discolouration in drinking water systems; the role of particles clarified*, IWA Publishing, London.

Vreeburg, J. H. G., Schaap, P. G., and van Dijk, J. C. (2004). "Measuring discoloration risk: resuspention potential method." Leading edge Technology conference, IWA, Prague.

Vreeburg, J. H. G. (2007). *Discolouration in drinking water systems: a particular approach.* Ph.D. thesis report.

Vreeburg, J. H. G., Blokker, E. J. M., Horst, P., and van Dijk, J. C. (2009). "Velocity based self cleaning residential drinking water distribution systems." *Water Science and Technology*, 9(6), 635–641, doi:10.2166/ws.2009.689.

VROM. (2005). *Insepctierichtlijn 5074: Harmonisatie Meetprogramma Drinkwaterkwaliteit* (Inspection directive 5074: Harmonisation measurement programme drinking water quality), RIVM, Bilthoven.

White, S., Milne, G., and Riedy, C. (2004). "End use analysis: issues and lessons." *Water Science and Technology: Water Supply*, 4(3), 57–65.

Yang, X., and Boccelli, D. L. (2009). "The impacts of demand variability on distribution system water quality." *CCWI 2009 'Integrating water systems'*, Sheffield, 459–463.

Appendix

Used statistics

A.1 AUTO- AND CROSS-CORRELATION

In Chapter 2, the auto- and cross-correlation are used to classify water demand patterns at different temporal and spatial scales. In Chapters 5 and 6, the auto- and cross-correlation of measured data and model results are compared to determine if the model results can be classified in the same way as the actual water demand patterns.

The (cross-) correlation coefficient is a quantity that gives the quality of a least squares fitting to the original data. Pearson's correlation factor ρ is defined as:

$$\rho = \frac{\sum_{i=1}^{N}(x_i - \bar{x})(y_i - \bar{y})}{\sqrt{\sum_{i=1}^{N}(x_i - \bar{x})^2}\sqrt{\sum_{i=1}^{N}(y_i - \bar{y})^2}} \tag{A.1}$$

Here x_i is the observed data, y_i is the model data; $\bar{x}$ is the mean of x_i and $\bar{y}$ is the mean of y_i. It can also be used to determine the correlation between two sets of observed data (x and w), as was done in this thesis (Chapter 6). In this thesis, R^2 was used to describe the similarity in shape between empirical and modelled distributions (Sec. A.1).

In case the datasets x and w are observed at the same location, but at different times, ρ is called auto-correlation. If the data are observed with a time step of 1 minute and the time difference between x and w is 10 minutes (i.e. $x(t) = w(t + 10)$ with t the time in min), ρ is called auto-correlation of lag 10. In case the data x and w are observed at the same time, but at different locations, ρ is called cross-correlation.

Auto-correlation is thus a measure of how predictable, and thus how variable, a dataset is over time. Water demand patterns from several days can be compared or the variability of the demand pattern over the day can be determined; when ρ is close to 0 it means a high variability for that time scale. A residential area may experience water demand patterns that are highly variable over the day (at a time scale of, for example, 1 minute ρ is equal to 0.2), but that are very predictable from historic water demand patterns (every Monday is the same, ρ is equal to 0.9). This shows that ρ by itself is not a complete measure for the variability of water demand patterns. In this thesis, the time scale for ρ is 15 minutes or less. It is used as a measure of variability of water demand in relative short time steps.

Cross-correlation is a measure of how much variation between two datasets of water demand patterns at different locations occurs. A residential area may be very similar to another residential area (ρ is close to 1), but it will have a different water demand pattern than an industrial area (ρ is much lower than 1).

Auto- and cross-correlation of water demand patterns vary with the temporal and spatial scale. Table A.1 shows the auto-correlation of two water demand patterns at one day and the next day at different spatial scales. At a large spatial scale (pumping station serving thousands of households) ρ is almost 1; at a lower level (a street of ca. 140 homes) ρ is much lower and at the single home level ρ is almost 0. At the single home there are a lot of data equal to 0; this is the reason that ρ is larger than 0.

In Chapter 4, the correlation coefficients of a set of 50 (simulated or measured) demand patterns are calculated, i.e. 1225 numbers in total, and represented as the mean $\pm$ the standard deviation (70% confidence interval).

A.2 GOODNESS-OF-FIT

The goodness-of-fit of a statistical model describes how well the model fits a set of measured data. Measures of goodness of fit typically summarise the discrepancy between observed values and the values expected under the model in question. In this thesis several types of statistical models are considered. Throughout the thesis, the SIMDEUM results (model) are compared to measurements such as flow patterns, flow characteristics and travel times. SIMDEUM is not a regression model. Therefore, a full regression analysis is not possible, but some of the goodness-of-fit parameters that are used in regression analysis were used here as well. In Chapter 3, the survey data is analysed and compared to specific probability distribution function (the model). Three goodness-of-fit parameters were determined, and next to that, a statistical test was performed to confirm the assumed probability distribution function (Sec. A.4).

The goodness-of-fit of the considered models was determined with three parameters, viz. two measures for the error between empirical and modelled distributions (Mean Error, ME; and Root Mean Square Error, RMSE) and a measure for the similarity in shape between empirical and modelled distributions (coefficient of determination, R^2). The coefficient of determination is the proportion of variability in a data set that is accounted for by the statistical model. In case R^2 equals 1, the variation is completely explained by the model:

$$ME = \frac{1}{N}\sum_{i=1}^{N} y_i - x_i \tag{A.2}$$

$$RMSE = \sqrt{\frac{1}{N-1}\sum_{i=1}^{N} (y_i - x_i)^2} \tag{A.3}$$

$$R^2 = 1 - \sum_{i=1}^{N} (y_i - x_i)^2 \Big/ \sum_{i=1}^{N} (x_i - \bar{x})^2 \tag{A.4}$$

Here x_i is the observed data, y_i is the model data and $\bar{x}$ is the mean of x_i. The ME and RMSE are expressed as absolute values and in percentages through comparison to the mean value of the measurement. The reason that R^2 (Eq. A.4) is used instead of ρ (Eq. A.1) is that for ρ the ME is removed from the fitting; ρ compares $x - \bar{x}$ to $y - \bar{y}$ and R^2 compares x to y. If a systematic error occurs (ME $\gg 0$), ρ can be close to 1, but R^2 will be small.

Table A.1 Auto-correlation (lag is one day) at different spatial scales

Spatial scale	Two measured water demand patterns	Auto-correlation
Pumping station (temporal scale 5 min)	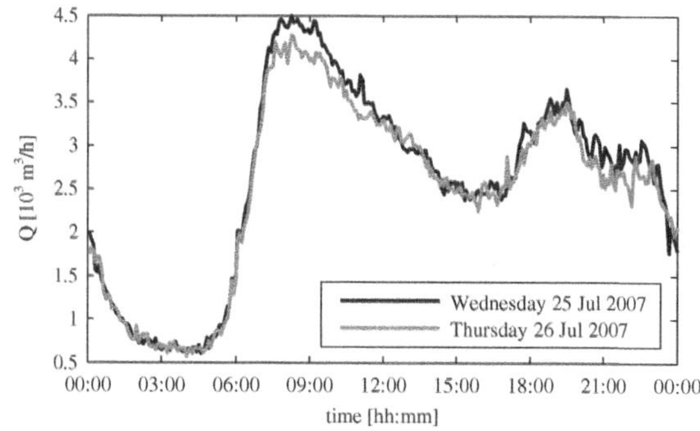	$\rho = 0.99$ 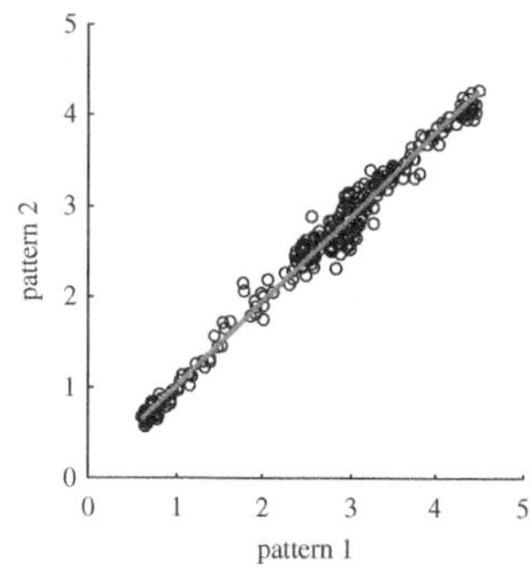
Street, 140 homes (temporal scale 5 min)	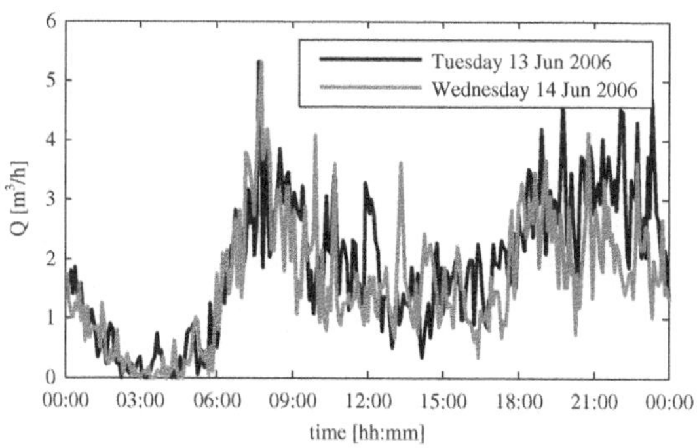	$\rho = 0.68$ 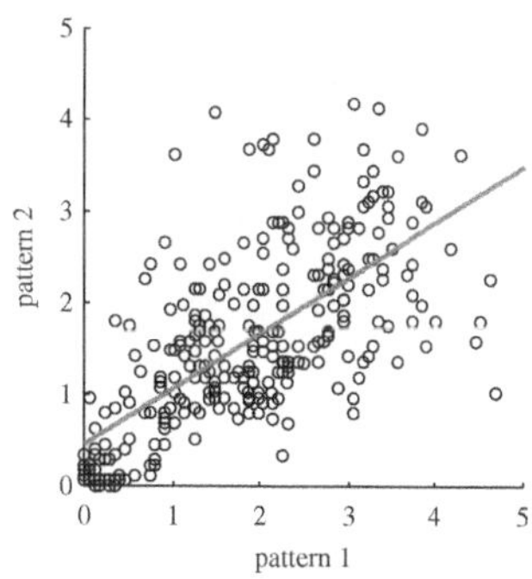
Single home (temporal scale 1 s)	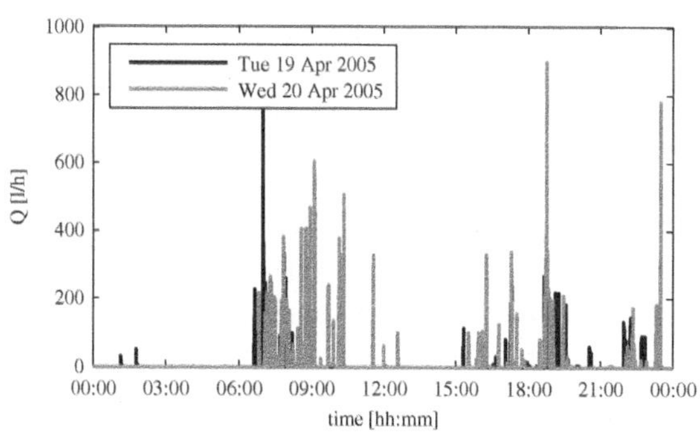	$\rho = 0.26$ 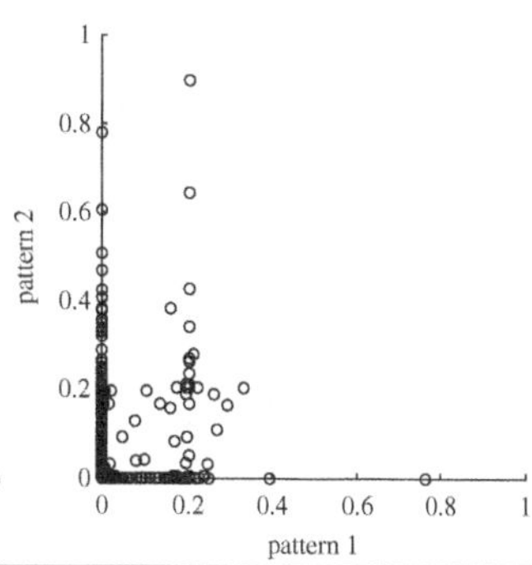

In the example of Figure A.1 there are measurements (x, the gray dots) and an underlying model (y, the black continuous line). The ME between data and model is the mean of the errors (gray bars) and is equal to −0.0045; the relative ME is −0.15%, i.e. the ME compared to the observed mean value (3.0045). The

RMSE is the square root of the means of the squared errors (black bars) and is equal to 0.1756; the relative RMSE is 5.84%, i.e. the RMSE compared to the observed mean value (3.0045). The R^2 of this example is equal to 0.95.

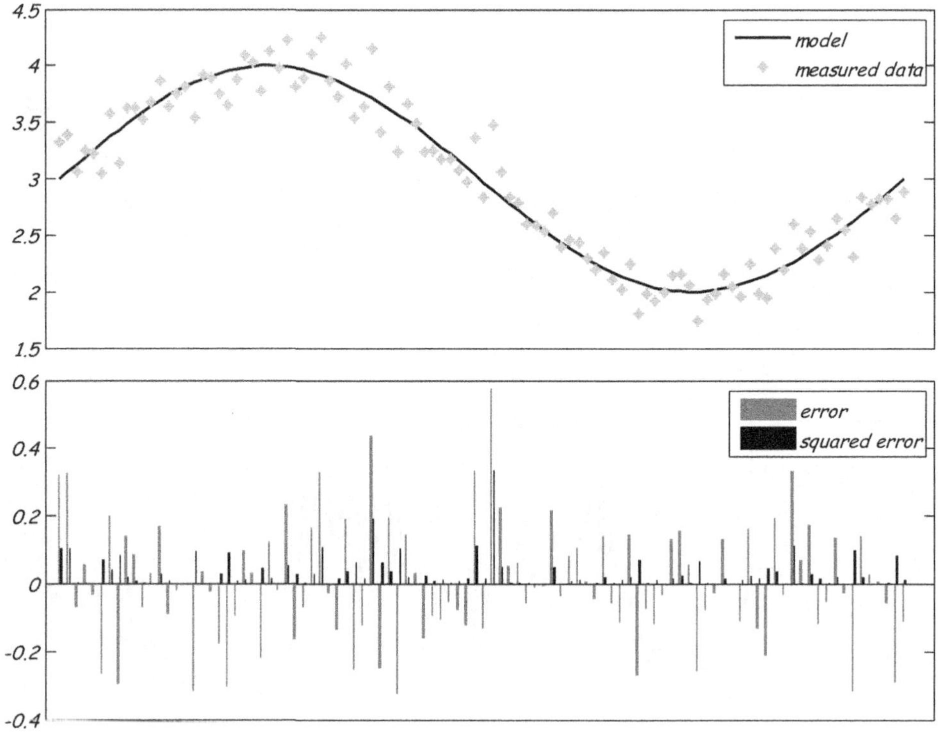

Figure A.1 Example of measurements, model and errors

The parameters Q_{max}, V, n_{pulse}, n_{hours}, Q_{CFD} and ΔQ_{CFD} are presented in the thesis as cumulative frequency distributions. The goodness-of-fit is expressed with ME, RMSE and R^2 for the cumulative frequency distributions. Q_{day} and C_d are patterns; the goodness-of-fit is expressed as ME and RMSE of Q_{day} and R^2 of C_d. Note that the ME and RMSE between the cumulative frequency distributions of the measured and modelled Q_{max} are not equal to the errors between the means of the measured and modelled Q_{max}. An example is shown in Figure A.2. In case the ME, RMSE and R^2 are compared of x and y by reference number n (Figure A.2a), the respective values are 0.18 (3.7%), 1.20 (23.9%) and <0 (correlation is 0). Because the occurrences of x and y are in random order it is not fair to compare them by reference number. Therefore, they are compared by their CDF (Figure A.2b). In this case, the goodness-of-fit values are determined for the CDF(x) and CDF(y) evaluated at the percentiles 1, 2, 3, etc. to 100; the respective values of ME, RMSE and R^2 are 0.19 (3.8%), 0.40 (7.9%) and 0.63.

In chapter 6, travel times were compared. The travel times are measured over several days which results in a variation of travel times over the day. The modelled travel times are the results of simulations with 10–14 different sets of demand patterns; this also results in a variation of travel times over the day. To compare measurements and model, an extra parameter was introduced, viz. the percentage of measured data points that deviate less than 10 minutes from what the model predicts. The accuracy of 10 minutes is arbitrary; it could also be 1 minute or an hour. The measure for an absolute accuracy allows that the measured data is not only compared to the average of the model outcome, but also to the 95% confidence interval.

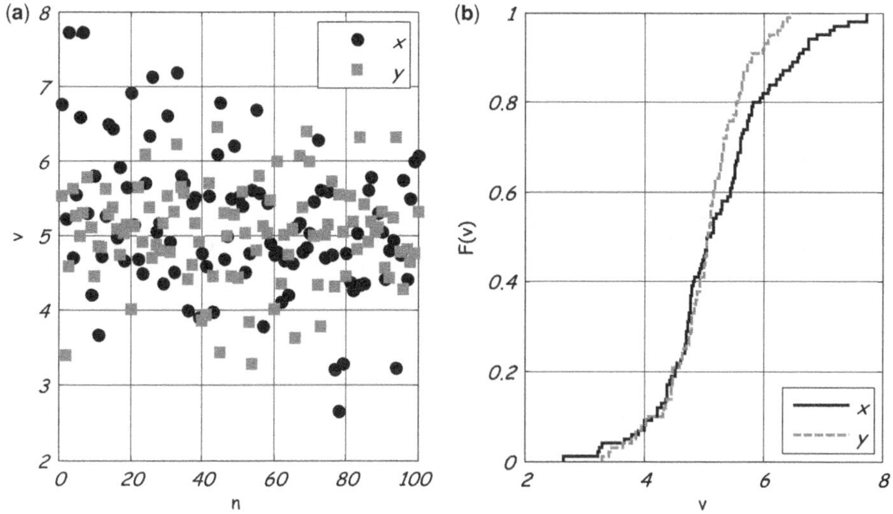

Figure A.2 Example of determining ME, RMSE and R^2 between *x* and *y* or CDF(*x*) and CDF(*y*)

An ME and RMSE of less than 25% and a R^2 of more than 0.7 are considered to be good. These values are not generally accepted in literature; every model can have its own thresholds.

A.3 PROBABILITY DISTRIBUTION FUNCTIONS

The probability distribution functions that are used in this thesis are summarized in Table A.2.

A.4 STATISTICAL TEST FOR NORMAL DISTRIBUTION

A.4.1 Introduction

There are several statistical tests available to determine if a dataset is normally distributed. Two were used in this thesis.

SIMDEUM simulates the sleep duration, time of getting up, time of leaving the house and the duration of being away from home based on the assumption that they are from a normal probability distribution. In Chapter 3, the data from the time-budget survey on these parameters were analyzed and a statistical test was used to determine if the data were normally distributed. Because these data are discrete values (multiples of quarters of an hour), a chi-square test was used.

In Chapter 6, 10 runs of network simulations with EPANET were done with 10 different sets of random water demand patterns. The resulting water age was represented as the mean water age and the 95 percentile confidence interval as derived from the mean and variance of the 10 calculated values for the water age. This approach is allowed if the data are normally distributed. To test that the data are normally distributed, the Lilliefors test was performed on the calculated water age over the day. This means that for each time step (96 time steps at the hydraulic time scale of 15 min) 10 data points were tested.

Table A.2 Probability distribution functions and their mean and variance

Distribution	Probability Distribution Function	Mean	Variance
Normal	$f(x\|\mu,\sigma)=\frac{1}{\sigma\sqrt{2\pi}}e^{\frac{-(x-\mu)}{2\sigma^2}}$	μ	σ^2
Lognormal	$f(x\|M,S)=\frac{1}{S\sqrt{2\pi}}e^{\frac{-(\ln x-M)}{2S^2}}$	$e^{(M+S^2/2)}$	$e^{(2M+S^2)}\left(e^{S^2}-1\right)$
χ^2	$f(x\|\nu)=\frac{x^{(\nu-2}e^{-x/2}}{2^{\nu/2}\Gamma(\nu/2)}$	ν	2ν
Poisson	$f(x\|\lambda)=\frac{\lambda^x}{x!}e^{-\lambda}I(0,1,...)^x$	λ	λ
Negative Binomial	$f(x\|r,p)=$ $\binom{r+x-1}{x}p^r(1-p)^x I(0,1,...)^x$	$\frac{r(1-p)}{p}$	$\frac{r(1-p)}{p^2}$
Beta distribution	$f(x\|a,b)=$ $\frac{1}{B(a,b)}x^{a-1}(1-x)^{b-1}$	$\frac{a}{a+b}$	$\frac{ab}{(a+b+1)(a+b)^2}$
Weibull distribution	$f(x\|a,b)=ba^{-b}x^{b-1}e^{-(x/a)}b$	$a\Gamma(1+1/b)$	$a^2\left(\Gamma(1+2/b)+\Gamma^2(1+1/b)\right)$

With Γ the gamma function and B the beta function.

A.4.2 Pearson's χ^2-test

A chi-square goodness-of-fit test tests the null hypothesis that the observed data are a random sample from a normal distribution with mean and variance estimated from the data, against the alternative that the data are not normally distributed with the estimated mean and variance.

The test is performed by grouping the data into bins (from $i=1$ to N), calculating the observed (O_i) and expected counts (E_i) for those bins, and computing the chi-square test statistic:

$$\chi^2=\sum_{i=1}^{N}(O_i-E_i)^2\Big/E_i \tag{A.5}$$

The quantity χ_s^2 is a measure of the deviation of a sample from expectation. Karl Pearson proved that the limiting distribution of χ_s^2 is a chi-squared distribution. The statistic has an approximate chi-square distribution when the counts are sufficiently large; each bin should contain at least 5 data points (MATLAB®). The probability that the distribution assumes a value of χ^2 greater than the measured value χ_s^2 is then given by (Weisstein):

$$P(\chi^2 \geq \chi_s^2) = \frac{\Gamma\left(\frac{N-1}{2}, \frac{1}{2}\chi_s^2\right)}{\Gamma\left(\frac{N-1}{2}\right)} \tag{A.6}$$

If the P-value exceeds the significance level (often 0.05) then the null hypothesis of a normal distribution is rejected.

A.4.3 Lilliefors test

The Kolmogorov–Smirnov test is a goodness-of-fit test for any statistical distribution. The test relies on the fact that the value of the sample cumulative density function is asymptotically normally distributed. In the Kolmogorov-Smirnov test, the cumulative frequency distribution CFD (normalized by the sample size) of the observations is calculated as a function of class. Then the CFD of a true distribution (most commonly, the normal distribution) is calculated. The greatest discrepancy between the observed and expected cumulative frequencies is called the "D-statistic". This is compared against the critical D-statistic for that sample size. If the calculated D-statistic is greater than the critical one, then the null hypothesis that the distribution is of the expected form is rejected (Weisstein).

The Lilliefors test is an adaptation of the Kolmogorov–Smirnov test. It is used to test the null hypothesis that data come from a normally distributed population, when the null hypothesis does not specify *which* normal distribution, i.e. does not specify the expected value and variance; instead the population mean and population variance are based on the data. Just as in the Kolmogorov–Smirnov test, the D-statistic is determined from the maximum discrepancy between the empirical distribution function and the cumulative distribution function (CDF) of the normal distribution with the estimated mean and variance. Finally, it is decided whether the maximum discrepancy is large enough to be statistically significant, thus requiring rejection of the null hypothesis. Here, this test differs from the Kolmogorov–Smirnov test. Since the hypothesized CDF has been moved closer to the data by estimation based on those data, the maximum discrepancy has been made smaller than it would have been if the null hypothesis had singled out just one normal distribution. To date, tables for this distribution have been computed only by Monte Carlo methods. The Lilliefors test was done with MATLAB® (Lilliefors 1967).

The Lilliefors test was performed on the resulting water age (10 data points) per time step (96 time steps at the hydraulic time scale of 15 min) on each location with the null hypothesis that the data are normally distributed. Figure A.3 shows for location 2 in the looped layout with a 15 min demand pattern time step the resulting water age of 10 runs at 96 time steps. Figure A.4 shows the Cumulative frequency distribution of 10 simulated water ages at 6:00 and 9:45 plus the corresponding normal distributions. The Lilliefors test did not reject the null hypothesis at the 5% significance level at 6:00, but did reject the null hypothesis at 9:45. This means that the data at 9:45 do not come from a normal distribution; at 6:00 the data are likely to be normally distributed. For the measurement location 2 in the looped layout with a 15 min demand pattern time step, the Lilliefors test showed that at 88 time steps (92% of 96 time steps) the null hypothesis could not be rejected at the 5% significance level. If the normal distribution can be confirmed for at least 90% of the time, it is assumed that the assumption of normal distribution is valid for the whole data range.

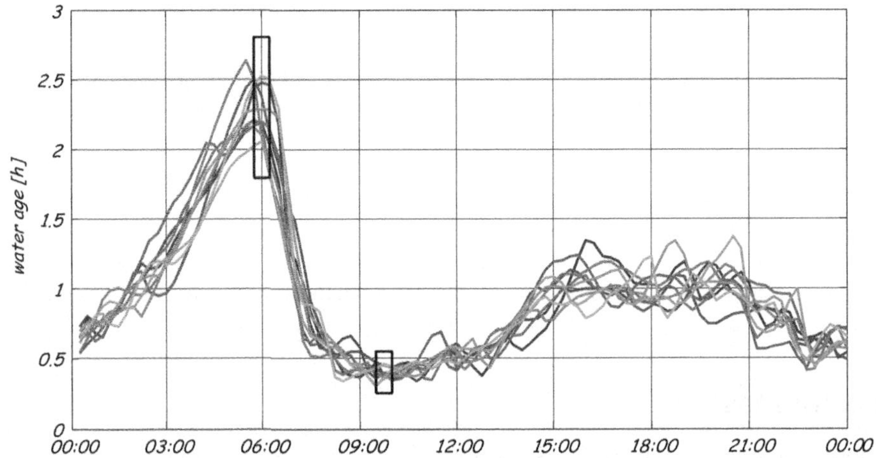

Figure A.3 Simulated water age (10 different runs) at measurement location 2 in the looped layout in Benthuizen (Chapter 6). The Lilliefors test is illustrated in Figure A.4 for time step 24 (6:00) and 39 (9:45) as indicated by the boxes

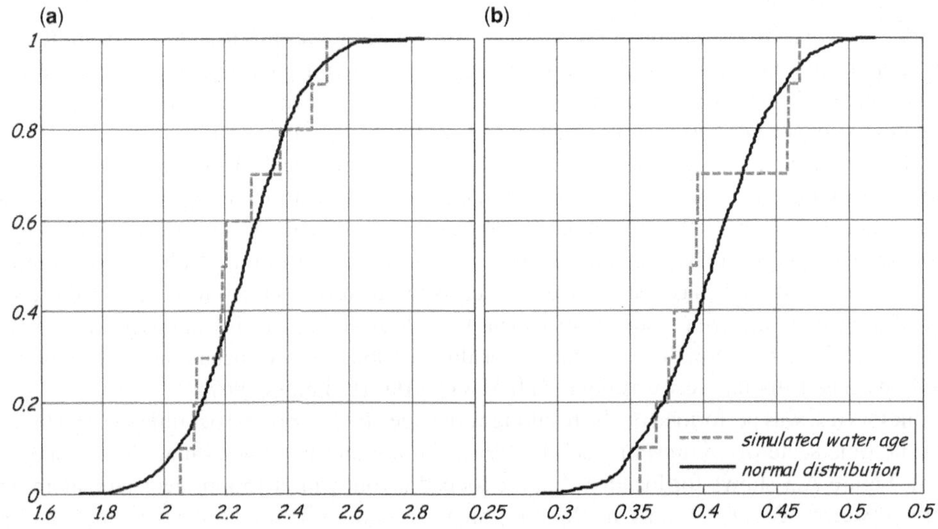

Figure A.4 CFD of simulated water age (10 different runs) and corresponding normal distribution at measurement location 2 in the looped layout in Benthuizen (Chapter 6) at time step a) 24 (6:00) and b) 39 (9:45). The Lilliefors test did not reject the null hypothesis at 6:00, but did reject the null hypothesis at 9:45

If the normal distribution is valid, it was assumed that the 95% confidence interval can be determined from the mean ± twice the standard deviation and that 10 model runs was enough to determine the 95% confidence interval.

REFERENCES

García, V., García-Bartual, R., Cabrera, E., Arregui, F., and García-Serra, J. (2004). "Stochastic model to evaluate residential water demands." *Journal of water resources planning and management*, 130(5), 386–394.

Lilliefors, H. W. (1967). "On the Kolmogorov-Smirnov test for normality with mean and variance unknown." *Journal of the American Statistical Association*, 62, 399–402.

Weisstein, E. W. http://mathworld.wolfram.com/Chi-SquaredTest.html, January 2010

Weisstein, E. W. http://mathworld.wolfram.com/Kolmogorov-SmirnovTest.html, January 2010

LIST OF SYMBOLS

Roman symbols

C	cross-sectional average particle concentration	mg/L
C_d	water demand multiplier pattern over the day	-
C_w	particle mass on pipe wall	mg/L
D	molecular diffusivity	m^2/s
D	pulse duration	s
E	mixing (axial dispersion) coefficient	m^2/s
E_T	Taylor dispersion rate	m^2/s
F	Frequency of use	1/day
I	pulse intensity	L/s
M	parameter of the lognormal distribution	-
Q	flow	L/s
Q_{CDF}	cumulative frequency distribution of flow over the day	L/s
Q_{max}	maximum flow per day	L/s
Q_{day}	flow over the day	L/s
Q_f	flushing flow	m^3/s
R^2	coefficient of determination	-
S	parameter of the lognormal distribution	-
T	Taylor time	-
V	Volume, total water use per day	L
d	pipe diameter	m
fl	reaction function	
h	index counting functional rooms	-
i	index counting busy times	-
j	index counting users	-
k	index counting end uses	-
l	pipe length	m
n	number of supplied homes	
n_{pulse}	number of pulses per day	-
n_{hours}	number of clock hours of water use per day	-

r	parameter of the negative binomial probability distribution	
p	parameter of the negative binomial probability distribution	
s	particle cloud height	-
t	time	s
t	PRP waiting time	s
u	mean velocity	m/s
u_d	particle threshold deposition velocity	m/s
u_{rs}	particle resuspension velocity	m/s
u_s	particle settling velocity, downward	m/s
v_{max}	maximum flow velocity	m/s
x	direction of flow	-
x_i	observed data	
y_i	model data	

Greek symbols

α	PRP mean pulse intensity	L/min
β	wall mass coefficient	-
β^2	PRP variance of pulse intensity	L/min^2
λ	parameter of the Poisson probability distribution	
λ	PRP mean number of arrivals per home	1/min
μ	mean of normal probability distribution	
ν	parameter of the χ^2 probability distribution	
π	user-specified multiplier pattern	-
ρ	correlation coefficient	-
σ	standard deviation of normal probability distribution	
τ	time at which tap is opened	hh:mm:ss
τ	PRP mean pulse duration	min
ω^2	PRP variance of pulse duration	min^2

Abbreviations

AC	Asbestos Cement
ADR	Advection-Dispersion-Reaction
AR	Advection-Reaction
BTO	Bedrijfstakonderzoek van de Nederlandse waterleidingbedrijvenJoint research programme of the Dutch water companies
BU	Bottom-Up
CI	Cast Iron
CFD	Cumulative Frequency Distribution
DMA	District Metered Area

DMP	Demand Multiplier Patterns
DWDS	Drinking water distribution systems
EC	Electrical Conductivity
FTU	Formazine Turbidity Unit
ME	Mean Error
MD	Maximum difference between $\mu + 2\sigma$ of 9th and 10th simulation run
LAM	Locally accumulated material
PDF	Probability Distribution Frequency
PVC	Poly Vinyl Chloride
PRP	Poisson Rectangular Pulse
RPM	Resuspension Potential Method
RMSE	Root Mean Square Error
SIMDEUM	Simulation of demand; an end-use model
TD	Top-Down